POWER
FROM
PLANTS

Research by the Energy and Environmental Programme is supported by generous contributions of finance and professional advice from the following organizations:

AEA Technology · Amerada Hess · Arthur D Little
Ashland Oil · British Coal · British Nuclear Fuels
British Petroleum · European Commission
Department of Trade and Industry · Eastern Electricity
Enterprise Oil · ENRON Europe · Exxon · Mobil
National Grid · National Power · Nuclear Electric
Overseas Development Administration · PowerGen
Saudi Aramco · Shell · Statoil · Texaco · Total
Tokyo Electric Power Company

POWER
FROM
PLANTS

THE GLOBAL IMPLICATIONS OF NEW TECHNOLOGIES
FOR ELECTRICITY FROM BIOMASS

WALT PATTERSON

THE ROYAL INSTITUTE OF
INTERNATIONAL AFFAIRS
Energy and Environmental Programme

EARTHSCAN
Earthscan Publications Ltd, London

AWH5271 -6/1

First published in Great Britain in 1994 by
Earthscan Publications Ltd, 120 Pentonville Road, London N1 9JN and
Royal Institute of International Affairs, 10 St James's Square, London SW1Y 4LE

Distributed in North America by
The Brookings Institution, 1775 Massachusetts Avenue NW,
Washington DC 20036-2188

A catalogue record for this book is available from the British Library.

ISBN 1 85383 208 1

The Royal Institute of International Affairs is an independent body which promotes
the rigorous study of international questions and does not express opinions of its
own. The opinions expressed in this publication are the responsibility of the author.

Earthscan Publications Limited is an editorially independent subsidiary of Kogan
Page Limited and publishes in association with the International Institute of
Environment and Development and the World Wide Fund for Nature.

Printed and bound by Biddles Limited, Guildford and King's Lynn
Cover by Elaine Marriott

Contents

Boxes

Acknowledgements

In the autumn of 1992 Silvan Robinson, then Chairman of the Energy and Environmental Programme at the Royal Institute of International Affairs, asked: 'Have you ever heard of biomass gasification?' I said that I had, and that it looked interesting. Silvan proposed that we explore it further; this report is the result. It would not have happened without Silvan's enthusiastic support from inception to completion, and it is dedicated to him with my grateful thanks. Silvan also introduced me to Roger Booth and Philip Elliott of the Shell International Petroleum Company, whose help and guidance have been invaluable in preparing the report, and in steering me around at least some of the many pitfalls in this fascinating but intricate subject.

Government and corporate staff people active in the field, including those from the Energy Technology Support Unit of the UK Department of Trade and Industry, the US Department of Energy, the European Commission and the European Parliament, have been unfailingly helpful in providing material and responding to queries. The study group for the draft report was especially useful, a knowledgeable and outspoken gathering of experts from the diverse fields involved, many of whom thereafter supplied additional material. My thanks to them all, and to Professor Jose Goldemberg, Dr Eric Larson, Gordon McKerron, Dr Peter Read and Dr Jeremy Woods, who examined the final draft meticulously and offered further assistance.

The Head of the Programme, Dr Michael Grubb, has kept a light but firm touch on the study throughout, and provided essential and incisive comment on its successive stages; I greatly appreciate his cordial but uncompromising supervision. Programme Administrator Matthew Tickle and Programme Assistant Nicole Dando have carried all the day-to-day responsibilities of organizing the project and seeing the study into print with unobtrusive, comprehensive and cheerful competence. To all my Programme colleagues, my thanks for making our work together both rewarding and enjoyable. Finally, to my wife Cleone all my love and thanks for keeping the home fires burning.

March 1994   Walt Patterson

Preface

Why should the Royal Institute of International Affairs conduct and publish a study on a currently minor energy source – focusing, furthermore, upon conversion technologies that have not yet been successfully demonstrated? The reason is that advanced technologies for power generation from biomass could have major implications stretching across national and sectoral boundaries. As the study shows, such technologies could become important in agriculture and land use policy, provide new avenues for existing power station engineering and gas turbine and water industries, and play a major role in strategies for carbon dioxide limitation – all in a context which could involve close 'North–South' linkages. Indeed, for many years these aspects may prove at least as important as any impact on energy balances.

These possibilities are all contingent upon successful development of complex technologies and systems. It would be hard to find anyone better suited to conducting such an assessment, examining the implications and then clearly communicating the results than Walt Patterson. His expertise in the closely related field of advanced coal-combustion power-generation technology, and his long experience of the international energy business, form an excellent background. We are also indebted to Philip Elliott and Roger Booth of the Shell International Petroleum Company, who are involved with the Brazilian biomass power project, for assisting Walt throughout.

This is Walt's first publication with the Energy and Environmental Programme since joining us as a Senior Research Fellow in the summer of 1993. He has been a great boost to our team, and, as this publication demonstrates, we are fortunate to have been able to enlist the capabilities of such an experienced, likeable and talented researcher.

March 1994

Dr Michael Grubb
Head, Energy and Environmental Programme

Executive Summary

Wood and other forms of biomass are exciting fresh interest in many parts of the world as fuels for generating electricity. The increasing availability and productivity of biomass fuels, and the development of innovative technologies to use them, promise to make so-called 'biomass power' an increasingly attractive option. Organizations now actively involved in developing biomass power include the European Commission, the US Department of Energy (DOE), many other national government departments and agencies, major engineering firms and utilities in Brazil, Finland, Sweden, the UK, the US and elsewhere, and the Global Environment Facility administered by the World Bank. The US DOE has declared that biomass power will be the most important renewable energy option for the next quarter-century, and has projected as much as 25 GW in operation in the US alone by 2010; the US Electric Power Research Institute has projected twice this amount.

By means of a range of technical innovations, including gasification and gas turbines, biomass power can generate the most versatile energy carrier, electricity, cleanly and efficiently from a 'renewable' fuel that can be stored and used as desired. Biomass not only absorbs as much carbon from the atmosphere as it emits when burned, but may do so for years before it is burned. Growing trees is therefore a way to mitigate climate change while establishing a biomass fuel resource. Since biomass power can also emit very low levels of nitrogen oxides and almost no sulphur, its impact on the atmosphere is very low, giving it a major environmental advantage, especially compared with coal-fired power.

Biomass power is already generated from agricultural and forestry residues; with new technologies such applications could be substantially expanded and made more economic, turning waste disposal problems into significant electricity supply opportunities. But a more substantial contribution to electricity supply would be based on 'energy crops' grown explicitly as fuel for power generation. In the EU, the US and elsewhere, attempts to withdraw land from food production to reduce the burden of surpluses are generating major political stress. Cultivating energy crops on such land would create an alternative economic activity that could bring new jobs and income to rural areas. Biomass power could give farmers and foresters a market for such energy crops.

The most suitable energy crops for biomass power, and likely yields, vary with climate, soil and other conditions. Current research is producing varieties that are more productive, that use water efficiently, that require minimal applications of chemicals and that are resistnat to pests and diseases. Agricultural research has historically not focused on the attributes of greatest importance to biomass energy, and the potential for improvement is high. However, growing energy crops could have significant and damaging environmental side-effects, including problems with monoculture, loss of biodiversity, soil degradation and possible overuse of chemicals. Environmental criteria for producing and using biomass fuel have already been devised and are being steadily upgraded; if biomass power is to be environmentally acceptable, they will have to be implemented with scrupulous care.

The range of technologies now becoming available may be able to generate electricity from units whose outputs range from a few tens of kilowatts up to 100 MW or even larger. Different technologies on different scales are also at different stages of development. Smaller free-standing units, perhaps using a biomass gasifier with a diesel or spark-ignition engine generator, could be used in rural areas to supply local power. Innovative technologies for industrial cogeneration from biomass fuel may allow biomass-based industries like sugar refineries to sell surplus electricity to local grids eager for extra capacity. Larger units, using one of the new range of high-throughput biomass gasifiers coupled to a gas turbine in a combined or other advanced cycle mode, could supply electricity to the grid. In some temperate latitudes units can also supply district heating. Biomass power has important site-specific aspects, including local growing conditions, local transport requirements for fuel, local heat loads and other factors. Optimal configurations are therefore complex and varied; the development of a range of technologies over many fronts will improve the scope and performance of different applications.

If biomass power achieves its potential, it will have a significant impact on atmospheric carbon dioxide, land use, agriculture, energy independence and security, coal markets and power plant markets. However, both biomass fuel production and biomass power generation still require a significant effort of research, development and demonstration before the advanced concepts necessary for widespread application reach the commercial stage. Some governments and international agencies are already providing financial support for RD&D for biomass power, but more vigorous promotion is warranted. In the long run, biomass power will have the biggest impact in tropical and subtropical regions, where biomass grows fastest and where the demand for electricity is also growing fastest. However, demonstrating economic and environmentally sustainable biomass power in industrial countries will help alleviate – and be driven by – environmental and agricultural problems. Success in industrial countries will foster the acceptance of biomass power as a valid option worldwide.

Introduction: Growing Interest

Can the oldest fuel make a comeback? Many now believe it can and should. The fuel in question is wood, along with other material from plants. This material is produced by biological processes that store the energy of sunlight in the substance of living plants. Such plant material is now called 'biomass'. People have been using biomass as fuel ever since our ancestors learned to start a fire with leaves and twigs. In many parts of the world, especially rural areas in developing countries, biomass is still the most important fuel – usually in the form of firewood, charcoal or animal dung, gathered for use with no commercial transaction involved. In industrial countries, however, biomass has long been displaced by fossil fuels – coal, oil and natural gas – and by hydroelectricity and nuclear power. In the 1990s, nevertheless, biomass may be on the threshold of a new breakthrough, as a fuel for advanced forms of electricity generation. As the concept excites mounting interest, it has even acquired a crisp new name: 'biomass power'. Indeed, it has already passed the purely conceptual stage. The first biomass power station using advanced technology is already being commissioned; more are on the way. Preliminary analyses suggest that biomass power, using the fuel and conversion technology best suited to the particular location and the particular application, could become a commercially competitive renewable energy supply option of major importance in many parts of the world, especially in tropical areas where biomass grows fastest. If biomass power, in its many possible manifestations, fulfils its potential, the global implications – for energy, agriculture, environment, land use, and rural and regional development – could be profound and far-reaching.

The idea of using biomass to generate electricity is not new. The US alone has some 6.5 GW (gigawatts or million kilowatts) of biomass power stations in operation burning biomass residues from agricultural and forestry, and about 2 GW more burning urban refuse.[1] But traditional stations like these, in the US and elsewhere, use conventional combustion technology that simply burns bulk biomass in a boiler. For some limited applications, in specific and favourable circumstances, this technology may be adequate; but for biomass in general such technology is likely to be inefficient and therefore uneconomic. The future of biomass power will be shaped by emerging technical innovations that promise to increase efficiency, lower costs and improve environmental performance. These technical innovations are being encouraged as a response to new pressures on agriculture and land use in industrial countries, and surging electricity demand in developing countries. The US Department of Energy (DOE) is now asserting that biomass power will be the most important renewable energy option for the next quarter-century.[2] Organizations now actively involved in biomass power include the US DOE, the European Commission, the US Electric Power Research Institute (EPRI), Royal Dutch/Shell Group companies, US General Electric, the Tennessee Valley Authority, the Finnish engineering firms Ahlstrom and Tampella, the Swedish engineering firm TPS Termiska Processer, the Swedish utilities Sydkraft and Vattenfall, the Finnish utility Imatran Voima Oy, the Brazilian utility Companhia Hidro Eletrica do São Francisco, the World Bank, the Global Environment Facility (GEF) administered by the World Bank, and a lengthening roster of other players.

Reasons for the growing interest in biomass power are easy to identify. According to the World Energy Council (WEC),[3] world energy use may increase from 8.8 Gtoe (gigatonnes or billion tonnes of oil equivalent) per year in 1990 to between 11.3 and 17.2 Gtoe in 2020 (see Figure 1.1, which shows four projections). Most of this increase will take place in what are still called developing countries. Within this total, moreover, the increase in electricity use will be yet greater; the WEC report's intermediate projection indicates an increase from 11,600 TWh (terawatt-hours or billion kilowatt-hours) in 1990 to 23,000 TWh in 2020 – effectively doubling in 25 years. Indeed, even in

---

[1] US Department of Energy, *Electricity from Biomass: National Biomass Power Program Five Year Plan*, Washington, DC: US Department of Energy, 1993.
[2] Ibid.
[3] World Energy Council, *Energy for Tomorrow's World*, London: World Energy Council, 1993.

Figure 1.1 Projections of energy demand and supply by 2020

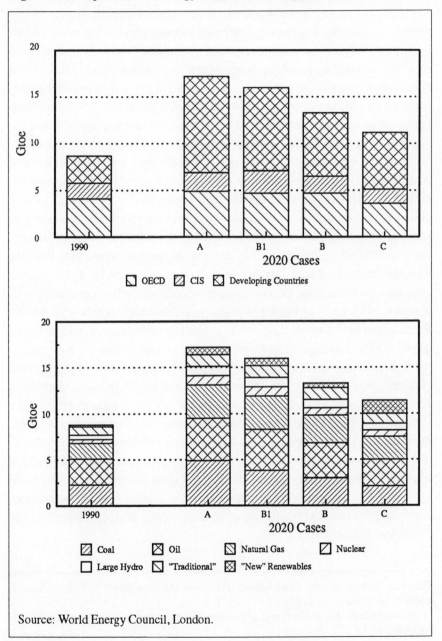

Source: World Energy Council, London.

industrial countries where total energy use may tend to stabilize, the use of electricity may continue to rise. What will provide the electricity to meet this increasing demand? Traditional generating technologies based on fossil fuels, hydroelectricity and nuclear power all face problems. Coal is cheap and abundant; but it emits more of the 'greenhouse' gas carbon dioxide per unit of energy than either oil or gas, as well as sulphur and nitrogen oxides causing 'acid rain'. Oil, because of its history of price volatility, has fallen from favour as a major fuel for baseload electricity generation. Natural gas is the cleanest fossil fuel; but in many parts of the world the necessary infrastructure may not be available for decades, and the longer-term availability and cost of natural gas may depend on transport, sometimes over thousands of kilometres, from politically unpredictable locations. Hydroelectricity is the classical renewable energy; but even where sites are still physically available (most sites in industrial countries are already developed), traditional hydroelectricity based on large dams has become intensely controversial, because it floods vast areas of land, displaces populations and upsets local ecosystems, and may also spread disease. Nuclear power emits no carbon dioxide from its reactors; but it faces problems of public acceptance, economic obstacles to financing in competition with other generating technologies, and stubborn questions about safety, waste disposal and the security of fissile materials.

The world therefore faces a choice of unsatisfactory options. At the United Nations Conference on Environment and Development in Rio de Janeiro in June 1992 almost all the governments of the world signed the Framework Convention on Climate Change, calling for controls on the emission of carbon dioxide.[4] They have agreed to a wide range of national and international controls on pollutants from fossil fuels.[5] They face public opposition to expansion of either hydroelectricity or nuclear power. Governments may be unable to meet their various undertakings without provoking bitter popular controversy. But failing to ensure a reliable and affordable supply of electricity will cause even more public unrest.

---

[4] Michael Grubb et al., *The Earth Summit Agreements*, London: Royal Institute of International Affairs/Earthscan, 1993.

[5] Caroline Thomas, *The Environment in International Relations*, London: Royal Institute of International Affairs, 1992.

Some believe that renewable energy in its wide range of manifestations will resolve the dilemma.[6] Renewable energy sources emit no net carbon dioxide; most emit no acid precursors or other pollutants. However, each has its limitations. Renewable energy resources are dispersed and of variable quality. Some – wind, solar thermal and photovoltaics – are intermittent, varying with the sunlight or the weather. Their economic status is still at best marginal, and in the case of photovoltaics significantly worse than that, except in niche markets. Wind power, in other respects the most advanced of the renewable energy technologies, is even meeting environmental opposition on the grounds of amenity and visual impact. All these renewable sources will certainly contribute, each in its appropriate context.[7] Nevertheless, of all the renewable energy technologies, biomass power possesses certain distinctive attributes that single it out for particular attention. Biomass power may be able to supply the most flexible and versatile energy carrier, electricity, from individual units ranging in output from less than one megawatt to tens of megawatts – in some places perhaps even more – with a steady and reliable output. Biomass releases its energy as heat; in suitable contexts biomass used for electricity generation can also supply heat, in the form of combined heat and power or cogeneration. The technology involved is already broadly familiar to electricity suppliers; the biomass fuel can be stored, is low in sulphur and – over a full cycle of growth and combustion – makes no net contribution to the greenhouse effect.

Perhaps the most persuasive immediate argument in favour of biomass power is that it may offer an imaginative and constructive approach to tackling one of the most pressing international problems: that of surplus agricultural production in industrial countries. The issue is complex, and subject to differing interpretations, as discussed in Chapter 2. But biomass power could offer an alternative economic activity in farming areas: growing biomass fuel. Biomass power may therefore attract support from two quite different lobbies in industrial countries, both already in existence and both influential: the power plant and turbine manufacturers, and the farmers. Demonstrating biomass power technology in industrial countries may also encourage its adoption in

[6] Thomas B. Johansson et al. (eds), *Renewable Energy: Sources for Fuels and Electricity*, London: Island Press/Earthscan, 1993.
[7] Ibid.

developing countries. Some developing countries, to be sure, will be better able than others to introduce biomass power; in localities where land or water is scarce and food production takes priority, biomass power may be inappropriate, whereas, in the many developing country localities where such constraints are less of a problem than electricity shortage, biomass power may prove a valuable option. In this study, all references to 'developing countries' should be understood to refer only to those areas where biomass power can be developed without conflicting with higher-priority demands on land or water. Because biomass grows so much faster in low-latitude developing countries, and because developing countries are especially eager to expand their electricity supply, biomass power could become a major economic and social contributor to sustainable development. As will be described in subsequent chapters, certain utilities, independent power producers and regional development authorities in developing countries are already putting their support behind biomass power as a potentially important element of their strategy.

Like the other renewable technologies mentioned above, biomass power still faces economic and environmental uncertainties. Two economic questions in particular remain to be answered: how will biomass power compare economically with other fuels and technologies, and how rapidly will new generating capacity be introduced? Many technical and institutional options can increase the efficiency of electricity use; improved efficiency will reduce the growth of electricity use and the need for additional supply technology of whatever kind. When new capacity is required, either to expand overall output or to replace capacity being retired, can biomass power hope to compete, for instance, with natural gas combined cycles, at present the most popular choice for new generating capacity? These questions will be explored further in subsequent chapters. Environmental questions, too, must be answered. Biomass in the form of residues is cheap close to the source; but its supply is limited. For biomass power to play a substantial role in electricity generation, power stations will have to have access to reliable supplies of biomass fuel grown for the purpose, in environmentally sound and sustainable ways. Biomass power issues thus intersect with agriculture and land use policies which are themselves already hotly controversial, not least environmentally. Biomass fuel production must not aggravate existing environmental problems, in either industrial or developing countries; indeed, it should be designed to alleviate them. To do so, however, it will have to meet stringent environmental criteria, and avoid the detrimental consequences of intensive cultivation and monoculture. These issues, too, will be explored further in subsequent chapters. If appropriate arrangements can be made, biomass power could help to alleviate some of the major problems now arising in this context, by offering a productive alternative use for agricultural land not required for food production in industrial countries. Moreover, because a biomass power station should optimally be sited close to the land on which its fuel is grown, biomass power could play a crucial role in rural and regional development, in both industrial and developing countries. A traditional fossil-fuelled power station, which can bring both technology and fuel from a long distance away, may look and operate much the same wherever it is sited; in a sense such technology is detached from its surroundings, exemplifying the concept of technology 'conquering nature', forcing it into a homogeneous mould. By contrast, biomass power is more 'site-specific': because the locality immediately around the power station supplies its fuel, the scale and type of technology used in the station must be suited to the local fuel supply, and thus to the characteristics of the locality – the land, the water, the climate and so on. In this respect biomass power, like other renewable energy technologies, exemplifies the concept of working with nature in all its diversity.

Indeed, the global dimensions of biomass power are varied and potentially substantial. Some international trade disputes arising from negotiations under the General Agreement on Tariffs and Trade (GATT) might be alleviated if surplus agricultural productive capacity could be used to supply fuel as well as food and fibre. Some companies are already planting forests – sometimes even in countries different from those in which their power stations are located – as a way to absorb fossil carbon, to offset that released by fossil-fired generating capacity. In the short term such offsets could help to ease the buildup of fossil carbon in the atmosphere, while creating a biomass resource that could subsequently be used to displace fossil-fired generation with biomass-fired generation. Because biomass fuel adds no net carbon dioxide to the atmosphere, a buildup of biomass power-generating capacity around the world could in due course contribute significantly to global efforts to mitigate emissions of greenhouse gases, thus furthering the aims of the Framework Convention on Climate Change. For this reason the Global Environment Facility administered by the World Bank has already begun to support biomass power development. Many developing countries seeking to expand their electricity systems may be able to use their own biomass resources for the purpose, provided that these resources are managed sustainably, and provided that the countries have access to the technology necessary to use their biomass efficiently and in an environmentally acceptable way. The World Bank may play a key role in this process. Moreover, in rural areas of developing countries, making modest amounts of electricity available locally for essential irrigation could boost local agricultural output of both food and energy crops; that is, appropriate biomass power technology could enhance the productivity of its own local fuel supply, while having valuable side-benefits for food production.[8]

This study endeavours to assess the current status and prospects of biomass power and to explore its implications. Chapter 2 describes the essential features of biomass, and the political context of biomass policy worldwide. Chapter 3 describes the concept of biomass power as an energy technology. Chapter 4 describes the activities already under way to develop biomass power in Sweden, the US, Brazil, Finland and elsewhere. Chapter 5 describes the considerations involved in designing an integrated biomass power station. Chapter 6 analyses the possible course of future development. Chapter 7 discusses the implications of the technical and economical success of biomass power. Chapter 8 offers conclusions and recommendations.

---

[8] Jeremy Woods and David O. Hall, *Bioenergy for Development: The Environmental and Technical Dimensions*, Rome: UN Food and Agriculture Organization, 1994.

Food, Fuel and the Future

2.1 Biomass: the resource

Photosynthesis stores solar energy in the substance of plants, producing energy-rich molecules including sugars, starches and lignocellulose. The energy can be recovered from these molecules by making them react with oxygen, directly or indirectly.[9] Plants also contain molecules of many other kinds, for instance proteins, oils and vitamins, created by the biological processes of growth. Their various constituent molecules make many different plants useful to humans. Since the dawn of humanity people have used plants for food, then for fuel, for timber and eventually also for fibre – for instance, cotton and paper-pulp. Only certain plants can be used for food, timber or fibre; but any green plant is potential fuel, because any green plant will burn, releasing the solar energy stored in its substance. A living plant, to be sure, contains a lot of water, and before it can be burned most of the water must be evaporated, to allow the temperature to get high enough to ignite the material – that is, the biomass must be dried, either in the fire itself or before the biomass is fed into the fire.[10] The energy content of biomass fuel is usually given in terms of dry weight; any remaining water content is a hindrance to combustion, although it may provide steam which is useful in certain applications, as will be described below.

This study considers biomass strictly in its role as an energy resource – that is, as a source of useful energy. Of course, burning is not the only chemical reaction that can release useful energy from biomass. Biochemical processes inside human and animal bodies release the energy of sugars and starches eaten as food; ruminant animals like cattle can also digest lignocellulose and use its energy. Biological processes like fermentation by yeasts have been

---

[9] David Hall et al., 'Biomass for Energy: Supply Prospects', in Johansson et al., *Renewable Energy*.
[10] Even so, the net 'calorific value' of biomass is positive up to some 88% water.

used for millenia to turn sugars and starches into ethanol, more commonly known as grain alcohol, the active principle in beers, wines and spirits, and a liquid fuel technically well suited to some spark-ignition engines. Fuel ethanol is now a major energy source in Brazil.[11] Still other biological processes that function in the absence of oxygen in biomass like animal or human dung produce a combustible mixture of methane and carbon dioxide. This so-called 'biogas' is produced in millions of 'digesters' in China and elsewhere, to supply light and heat and even drive internal combustion engines for electricity generation and other uses in rural areas.[12] Moreover, some plants, for instance oilseed rape, produce natural oils that can be processed into liquid fuels, albeit at a cost currently high compared to that of petroleum products.[13] This study, however, will focus on the processes that can use the energy content of biomass to generate electricity.

Historically, people have cultivated plants known to be good sources of food, timber and fibre, and studied such plants intensively. Comparatively little work has been done, however, on the cultivation of plants as good sources of fuel. Most of the firewood used in the 1990s, especially for cooking in developing countries, is still gathered from woodlands or forests growing more or less naturally. The firewood is harvested, but not cultivated; its replacement is left to nature, and all too often nature is losing the battle.[14] By contrast, most of the biomass fuel now used industrially is material left over when cultivated plants are processed for timber, fibre or food: wood shavings and sawdust from sawmills, bark from papermills, rice hulls, nutshells, bagasse from sugar cane and so on .[15] As fuel, this biomass is an afterthought, although often a worthwhile one. Because it is an afterthought, and originally a waste material, it is essentially free, and indeed may pose a disposal problem. Accordingly, so long as it need not be transported any distance, it can be

---

[11] Jose Goldemberg et al., 'The Brazilian Fuel Alcohol Program', in Johansson et al., *Renewable Energy*.

[12] P. Rajabapaiah et al., 'Biogas Electricity: The Pura Village Case Study', in Johansson et al., *Renewable Energy*.

[13] European Commission, DGXVII, *The European Renewable Energy Study: ESD Ltd, DECON GmbH, ITC Ltd and ERL*, Brussels: European Commission, 1994.

[14] The fuelwood issue, however, is more complex than once believed; see e.g. *Renewable Energy for Development* (Stockholm Environment Institute), Vol. 6, No. 2, December 1993.

[15] Hall et al., 'Biomass for Energy'.

burned inefficiently but more or less economically in a small traditional boiler on the processing site. The only cost of the heat or electricity is the capital and maintenance cost of the boiler, plus perhaps a modest expenditure for fuel handling; the fuel itself is free. Such arrangements, however, useful though they may be, are necessarily limited to sites that have ready access to such biomass residues. They cannot contribute more than marginally to the electricity requirements of a whole country. If biomass power is to be a truly significant option for electricity supply, the entire generating system – fuel supply, transport and processing, generating plant and waste management facilities – must be an integrated, reliable and sustainable system, conceived, developed and operated as such. It may utilize either biomass residues or purpose-grown biomass fuel; but even a system based on residue fuels will benefit from planning that encompasses the residues not as 'waste' but as useful byproducts from the outset. To envisage such a system, and how it might be realized, consider first its most distinctive feature: growing plants explicitly to be burned as fuel.

2.2 Agricultural dilemmas and pressures

Growing plants for a purpose – that is, agriculture – has long been recognized as the basis of a stable human society. At its most basic it is an uncomplicated if arduous process: a family lives on a plot of land on which it cultivates a variety of edible plants to feed itself, and possibly a few domestic animals. If it can grow more than it needs for its own sustenance, it can use the surplus to trade with its neighbours. In rural areas of developing countries agriculture still conforms to this basic pattern. Elsewhere, however, it has evolved almost beyond recognition. By using machinery for planting and harvesting, adding artificial nutrients to the soil and controlling pests and diseases with chemical remedies, a single person can cultivate many hundreds of hectares and gather a vastly increased harvest from the same area of land. Moreover, selective breeding has created plant varieties that grow larger and faster, crop more heavily and are more resistant to pests and diseases. Whether such patterns of agriculture are environmentally sustainable, however, is now a matter for some concern; and in any case they already have an awkward corollary. Major agricultural areas like the US, Canada and the EU can now grow far more food than their citizens need. If you produce more food than your customers want to buy, the price of the food falls, and so does your income. If you try to produce more food to boost your income, the price falls further, in a vicious circle.

Governments do not want farms to fail and food production to cease; moreover, farmers in industrial countries represent a potent political force. Accordingly, faced with this dilemma, national governments have implemented a sprawling bureaucratic superstructure to control and direct agricultural production, pricing and trade, both nationally and internationally. Its most famous – or notorious – manifestation is the Common Agricultural Policy (CAP) of the EU.[16] The CAP emerged during the 1960s to support European farmers and encourage them to increase their output, to make the original six members of the then European Economic Community self-sufficient in food. It succeeded all too well. Overproduction of grains, meats, dairy products and other foodstuffs has created 'mountains' of surplus products purchased and stored by EU governments to maintain prices. Subsidies paid to farmers now constitute much the largest single item of the EU budget. In the 1990s the EU is grappling with a fundamental mismatch between its potential food production from the available cultivated land and the amount of food that EU citizens can eat. Protracted and tortuous negotiations led eventually to an agreement in 1992 on reform of the CAP.[17] A key feature of the reform was the decision to make CAP production subsidies available only to farmers prepared to 'set aside' each year a designated fraction – 15%, measured according to stipulated criteria – of their potentially productive land, and to grow no food on it. How this 'set aside' policy will work in practice remains to be seen. The US Department of Agriculture has operated a similar policy within the US for many years, without satisfying either farmers or consumers. Farmers do not like to leave formerly productive land uncultivated; it is frustrating and contrary to their instincts, and they are concerned that such land may also become a breeding ground for weeds, pests and diseases that

---

[16] See e.g. Ulrich Koestler and Malcolm D. Bale, 'The Common Agricultural Policy: A Review of its Operation and Effects on Developing Countries', *World Bank Research Observer*, Vol. 5, No. 1, January 1990.
[17] J. M. Rollo, 'Reform of the Common Agricultural Policy', *The World Today*, Royal Institute of International Affairs, London: January 1992.

can affect adjoining cultivated land. Some environmentalists dispute this view. They argue that modern intensive farming methods are the real problem, and that so-called 'extensification' – reducing the use of chemical nutrients, pesticides and herbicides and otherwise adopting more so-called 'organic' methods – would reduce stress on land and habitat and make agriculture more sustainable, while lowering surplus production. They also argue that land withdrawn from cultivation enhances biodiversity, supporting a wider variety of plant and animal species than cultivated land. Others ask how farmers are to earn a living if they do not farm their land and sell what it produces.

Agricultural policy has created not only intense controversy within the EU, but also fierce tension between the EU and its trading partners – notably the US, not backward itself in the matter of agricultural subsidies. The US wants access to EU markets for its own agricultural produce; it also deeply resents subsidized exports from the EU that undercut US exports to markets outside the EU. Developing countries, many of which would export food as a key way to earn foreign exchange, suffer still more. The agricultural policies of major producers like the EU and the US have led into an international minefield. A guiding principle of GATT, established after the Second World War as a basis for global trade between nations, has always been removing barriers to international trade. The transformation of the global marketplace since the inception of GATT in due course prompted discussions called the Uruguay Round, initiated in 1985, to reorganize GATT accordingly. To do so took some eight years; agreement on the Uruguay Round was reached only in December 1993. The discussions encountered many disparate difficulties, but one of the most intractable arose from the agricultural policies of bodies like the US and the EU. Other countries, including many developing countries as well as the emerging democracies of eastern Europe, now produce food commodities they wish to export to the US, the EU and other OECD countries. They could sell their food to these countries at low prices and still make a profit in hard currency, crucial for their economic development. But farmers in the US and the EU, already struggling against the consequences of domestic food surpluses, do not want to have to cut their prices to meet competition from low-priced imports. Giving developing countries and eastern Europe access to the markets of the OECD to sell their food would be one of the most effective forms of development assistance available; but it would gravely aggravate the problems of agriculture within OECD countries.

This study can only sketch the outlines of this daunting global dilemma. That dilemma, however, adds a fresh urgency to consideration of one possible alternative option. Suppose the land no longer required to grow food were to be used instead to grow fuel? If the fuel could be sold at an appropriate price the land would remain productive, and provide an income to the farmer. Initially, subsidies previously paid simply to keep the land out of production could be reallocated to support fuel production. Subsequently, as the technology of growing and converting fuel crops developed, these subsidies could be reduced and perhaps, ultimately, eliminated. Problems will be inevitable; but the option nevertheless offers obvious attractions – so much so that it is already under intensive investigation by policymakers in the US, the EU and elsewhere.

2.3 Biomass energy

One avenue of approach, already extensively canvassed as noted above, has been to consider the prospect of manufacturing liquid fuels – particularly ethanol, methanol and rape methyl ester (so-called 'biodiesel') – from crops grown for the purpose. But another approach may offer more immediate promise: growing crops to fuel electricity generation – that is, biomass power. Per unit of energy, electricity attracts a significantly higher price than liquid fuels.[18] Moreover, if the concept of biomass power based on energy crops and advanced technology can be established in industrial countries, it may become even more important in developing countries, as will be discussed below.

In its simplest form, of course, biomass power does not depend on energy crops. All the biomass power stations now in operation use biomass residues, from agriculture and forestry and urban refuse, and other residues from processing timber, food and fibre (see Tables 2.1 and 2.2). This approach to biomass power can certainly be applied more widely in appropriate circumstances. But it has serious limitations. Even if the fuel is free at its source, the

---

[18] Philip Elliott and Roger Booth, *Brazilian Biomass Power Demonstration Project*, Special Project Brief, Shell, London, 1993.

Table 2.1: Energy content of selected biomass residues: exajoules per year

| | | | | | | Roundwood | |
| | | | | | | | Fuelwood/ |
Region	Maize	Wheat	Rice	Sugarcane	Dung	Industrial	charcoal
Industrialized							
US/Canada	2.95	1.93	0.13	0.19	3.08	7.66	0.92
Europe	0.61	2.39	0.04	0	4.22	4.12	0.41
Japan	0	0.02	0.24	0.01	0.30	0.41	0
Australia	0	0.29	0.02	0.19	1.36	0.35	0.02
New Zealand							
Former USSR	0.23	1.97	0.04	0	3.58	3.92	0.60
Subtotals	3.8	6.6	0.5	0.4	12.5	16.5	1.9
Developing							
Latin America	0.71	0.38	0.29	3.58	7.21	1.47	2.12
Africa	0.48	0.25	0.20	0.54	5.38	0.75	3.31
China	1.23	1.75	3.43	0.48	4.81	1.27	1.34
Other Asia	0.51	1.88	5.29	2.70	10.91	2.31	4.62
Oceania	0	0	0	0.03	0.02	0.05	0.04
Subtotals	2.9	4.3	9.2	7.3	28.3	5.8	11.4
World	**6.7**	**10.9**	**9.7**	**7.7**	**40.8**	**22.3**	**13.3**

Source: *Renewable Energy: Sources for Fuels and Electricity*, Thomas B. Johansson et al., Island Press/ Earthscan, 1993.

supply of residue fuel that can be delivered to a plant without incurring substantial costs for transport means that in general a plant of this type is small – at most a few megawatts, and often smaller still. (Moreover, with the exception of municipal refuse, biomass residues are seldom in practice free; even if the residues have a zero or negative value to their owner, when a fuel supply contract is being negotiated the owner knows that the material will acquire value when the power station is built, and will want a share.) The residue fuel is simply burned in a conventional boiler to raise steam; and the small size of the unit imposes inherent inefficiencies on a steam-cycle system. The conversion efficiency for electricity generation will be less than 25%,

Table 2.2: Electricity generating plants burning biomass fuels in the US

	Number of facilities		Installed capacity		
State	Stand alone	Cogeneration	Stand alone	Cogeneration	Total
Alabama	0	15	0	375	375
Arizona	2	0	45	0	45
Arkansas	1	4	2.4	10	12
California	64	30	736	255	991
Connecticut	4	3	155	14	169
Delaware	1	0	13	0	13
Florida	12	15	314	474	788
Georgia	0	5	0	36	36
Hawaii	2	13	70	129	199
Idaho	1	6	0.2	116	116
Illinois	0	1	0	2	2
Indiana	0	7	0	36	36
Iowa	2	1	11	2.2	13
Kentucky	1	1	1	1	2
Louisiana	1	12	11	300	311
Maine	4	22	88	704	792
Maryland	2	2	214	94	308
Massachussetts	2	9	38	252	290
Michigan	3	13	78	247	325
Minnesota	3	23	63	161	224
Mississippi	0	10	0	230	230
Missouri	0	2	0	60	60
Montana	2	17	18	340	358
New Hampshire	3	5	15	65	80
New Jersey	2	0	14	0	14
New York	11	17	154	425	579
North Carolina	3	27	60	351	411
Ohio	1	6	17	90	107
Oklahoma	2	1	8	17	25
Oregon	3	24	69	185	254
Pennsylvania	0	9	0	144	144
South Carolina	1	13	49	46	95
Tennessee	2	12	6	43	49
Texas	1	9	2	146	148
Utah	0	1	0	20	20
Vermont	5	3	80	218	296
Virginia	0	9	0	136	136
Washington	3	11	72	120	192
Wisconsin	5	9	55	117	172
Total	**149**	**367**	**2,459**	**5,962**	**8,421**

Source: *Renewable Energy: Sources for Fuels and Electricity*, Thomas B. Johansson et al., Island Press/ Earthscan, 1993 (based on *National Biomass Facilities Directory*, National World Energy Association, Arlington VA, 1990)

often much less. Furthermore, the unit capital cost of such a small steam-cycle plant is high. These various constraints make biomass power based on residue fuels and steam-cycle technology at best only marginally economic, except in particular industrial contexts, for instance on the site of a sawmill, a pulp mill, a sugar factory or similar installation, or as a way to dispose of the combustible fraction of urban refuse, with electricity as a byproduct. (A related approach is to utilize the combustible gas that accumulates in landfill sites as refuse decays; such landfill gas is already essentially economic where it is available, and is now coming into use to generate electricity in the UK and elsewhere. But the resource is limited, and landfill itself is falling out of favour as a waste disposal option. Certainly no one wanting to recover useful energy from urban refuse would start by creating a landfill.) Such applications can certainly be expanded significantly; but the constraint of transport costs will always be a limiting factor, possibly aggravated by a highly dispersed resource and low conversion efficiency.

If biomass power is to make a major contribution to electricity supply, then a different approach, both to fuel supply and to the technology to use it, is required. An integrated biomass power station, built on its own, rather than as an adjunct to some other industrial process, to generate electricity or combined heat and power, may have an output in the range from a few hundred kilowatts to tens of megawatts; and it will require a reliable, substantial supply of biomass fuel, grown on a sustainable and environmentally acceptable basis, either as a dedicated sustainable residue – perhaps more accurately a byproduct – or as a dedicated energy crop. Because this fuel will be more expensive to grow, to transport and to process than industrial process waste, it must be utilized more efficiently, to generate enough marketable electricity from each tonne of fuel. That in turn will require technology more advanced than the traditional boiler and steam turbine-alternator. Fortunately, both these issues – biomass fuel supply and biomass fuel conversion technology – are now the focus of major research and development activities in a number of countries, with financial support from industry, government and international organizations, as will be described in the next two chapters.

Biomass power involves a constituency that brings together an unusual cluster of otherwise disparate interests, including among others electricity suppliers and users; farmers and foresters; engineering equipment manufacturers; agricultural and forestry equipment manufacturers; energy, agricultural, forestry, trade and environmental departments of governments and international agencies; and regional and development agencies. One important group whose interest in biomass power might not seem obvious is water companies. In many industrial countries, for instance the UK, water companies are facing the problem of disposing of ever-increasing accumulations of sludge from sewage treatment plants. This sludge is rich in nutrients that could be valuable to crop plants; but because the sludge may also contain industrial contaminants its use on food crops is restricted or even banned. No such restriction, however, except for controls on some heavy metals, would prevent the use of sewage sludge as a nutrient for energy crops; indeed, the nutrients would boost the productivity of the energy crops. Moreover, the roots of some energy crops appear to retain nutrients more effectively, reducing run-off into water supplies. Indeed, some developing countries that may not as yet have sewage treatment plants might also benefit from disposing of sewage onto non-food plantations like energy crops, simultaneously reducing pollution and accelerating the growth of the plantations.

Biomass power could therefore offer advantages to many different potential beneficiaries. Nevertheless, many hurdles remain to be crossed. For a start, many of the prospective beneficiaries of biomass power still know little or nothing about it. Because biomass power is so site-specific, its advantages and disadvantages will be different in different locations, as will the incentives and impediments affecting its progress. Each biomass power proposal will have to be evaluated in context; generalizations may be misleading or simply wrong. Nor will biomass power be without its opponents. The world coal industry and the world nuclear industry, to name but two, will not welcome additional competition. The attitude of environmental groups may be ambivalent. Biomass power is based on a renewable energy resource, with certain clear environmental advantages over traditional electricity supply technologies. Nevertheless, biomass power could have major impacts on land and water use and quality, wildlife habitat and biodiversity, in both industrial and developing countries, unless it complies with strict environmental criteria. Accordingly, despite its potential and its wide constituency, biomass power has a long way to go. Just how far, and what that may imply for various interest groups, for and against, is the subject of this study.

Biomass Power: The Basics

3.1 Biomass as a fuel

If you leave appropriate biomass for a few million years under suitable temperatures and pressures, it becomes coal. Because they are thus in a sense first cousins, biomass and coal have a lot in common. Both are solid fuels. Both vary over a wide range in composition, physical and chemical characteristics, and energy content. Both have to be processed appropriately before use – in particular, both must be dried and reduced in unit size to acceptable dimensions to feed into a boiler or other energy conversion unit, which requires in turn a fuel-handling system adequately robust to cope with solids without clogging or jamming. In general, coal has a somewhat higher energy density than biomass, perhaps 26 GJ (gigajoules) per tonne for a bituminous steam coal compared to perhaps 19 GJ per tonne for oven-dried biomass.[19] At the point of production, however, raw biomass contains a lot of water, making its energy density much lower than this. Since shipping material of such low energy density a significant distance is disproportionately costly, biomass must be either dried at the point of production or used close to the point of production – ideally both.

On the other hand, unlike coal, biomass contains very little sulphur – 0.04% by weight might be a typical number – or heavy metal, making both its flue gas and its ash less pernicious than those of coal. Indeed, biomass contains little ash – about 0.5–5% by weight. On a dry ash-free basis, biomass tends to have a chemical composition of about 50% carbon, 6% hydrogen and 44% oxygen.[20] Biomass also tends to be over 80% 'volatiles' – breakdown products that are driven off in gaseous and vapour form when biomass is heated.

---

[19] Expressed in terms of 'lower heating value' (LHV): that is, net of the energy that escapes in the steam formed by burning the hydrogen of the fuel; this energy would be recovered only if the steam were condensed to water.

[20] Robert H. Williams and Eric Larson, 'Advanced Gasification-based Biomass Power Generation', in Johansson et al., *Renewable Energy*.

The volatiles include combustible gases like methane and other light hydro-carbons, tar vapours, hydrogen and carbon monoxide, together with a smaller proportion of non-combustible gases like water vapour and carbon dioxide. Coal, by comparison, is usually less than 50% volatiles, often substantially less. Simply heating biomass drives off these volatiles rapidly – within seconds – leaving a highly reactive solid 'char'. Biomass is therefore more reactive than coal, making its energy content easier to mobilize. However, because biomass is in other ways so similar to coal, technologies that can use coal can often be adapted to biomass, generally with little fundamental modification; while the differences between biomass and coal may be crucial to reducing the capital cost of such technologies for use with biomass.

Comparing biomass and coal shows that each has advantages and disadvantages. At the moment, internationally traded coal is likely to be cheaper per unit of energy content than purpose-grown biomass, depending on transport costs – although some EU coal, for instance, is heavily subsidized. Coal is less bulky than biomass and is easier and more convenient to handle and process; and coal production requires much less land than biomass production. Biomass has a lower energy density than coal. On the other hand, two further differences, both fundamental, favour biomass. Mobilizing the energy from coal releases into the atmosphere fossil carbon that was extracted from the atmosphere by photosynthesis millions of years ago. This additional carbon dioxide reinforces the greenhouse effect. By contrast, mobilizing the energy from biomass merely returns again to the atmosphere the carbon that has recently been extracted from it by the growing plants. In other words, although some net carbon dioxide emissions will arise from the fuels used to grow, harvest and transport the biomass, the energy from biomass fuel itself makes no net contribution to the greenhouse effect. Moreover, once coal at a given location has been extracted and used it is gone. Under the right conditions, however, a given location can continue supplying biomass indefinitely, as plants still growing replace those harvested for use.

3.2 Producing biomass

Accordingly, the essential prerequisite for successful use of biomass fuel – whether as dedicated-residue byproducts or as purpose-grown energy crops –

is to establish and maintain at a given location the right conditions for plants to grow with healthy vigour. Relevant factors are always specific to a given location, and some are essentially predetermined. They include:

- the intensity and duration of sunlight through the year;
- rainfall or other water supply;
- the annual temperature profile;
- soil condition;
- natural nutrient availability; and
- the prevalence of pests and diseases.

In arable areas of developed countries all of these factors have been studied and recorded for many years. Data about developing countries may be neither as complete nor as reliable; improving the quality of these data will be a necessary first stage if biomass power is to contribute effectively to electricity supply in developing countries.

Depending on the attributes of a given location, a variety of different plants may be suitable fuel sources for biomass power. The most promising plants include 'herbaceous' plants such as fast-growing annual and perennial grasses, and fast-growing 'woody' plants – that is, trees. In principle any biomass whatever might be a fuel source for biomass power, since all biomass is combustible. However, if crops are to be grown expressly as fuel for an economic activity, the factor of overriding importance is productivity: how much biomass can be produced on a continuous, sustainable and environmentally acceptable basis from a given area of land. This will depend not only on the local growing conditions, but upon how particular plant species respond to these conditions – that is, how fast they grow. In temperate latitudes the woody species – trees – of most interest at the moment are varieties of conifers like fir, spruce and pine, well known to conventional forestry, and fast-growing varieties of poplar and willow. According to the European Commission, at present conifers can, under optimal conditions, yield 5–6 oven-dried tonnes per hectare per year (odt/ha-y), willow about 10 odt/ha-y and poplar about 12 odt/ha-y.[21] The same source estimates future productivity rising to 8–10

---

[21] G. Grassi and T. Bridgewater (eds), *Biomass for Energy and Environment, Agriculture and Industry in Europe*, Brussels: European Commission, 1992.

odt/ha-y for conifers, 16 odt/ha-y for poplar and 17 odt/ha-y for willow. Whatever the species, achieving sustainably high yields will always require good management. Field trials have been under way in many places for more than a decade, to assess the performance of different varieties and different 'clones' – genetically identical plants propagated by cuttings from a single parent plant.[22] In southern Europe varieties of robinia, cynara and eucalyptus look more promising; the European Commission gives current yields from eucalyptus as 10–15 odt/ha-y, rising in the future to 17 odt/ha-y. In tropical and sub-tropical latitudes the trees of most interest at the moment include eucalyptus and pine. The grasses of interest in temperate latitudes include so-called 'elephant grass' or miscanthus in northern Europe and 'switchgrass' in the US, sweet sorghum in southern Europe and – much the best-known of the energy-crop grasses – sugar cane throughout the tropics and sub-tropics. Sugar cane is one of a limited group of plants in which photosynthesis produces a type of sugar with not three but four carbon atoms – a so-called C4 plant. C4 plants grow much faster than the more numerous C3 plants – all trees and most other plants – in tropical conditions. In 1987 the worldwide average yield of dry matter above ground for sugar cane was 36 tonnes of dry matter per hectare per year; the highest yield over a whole country was achieved in Zambia, with 77 tonnes of dry matter per hectare per year.[23] Sugar cane, however, tends to require high levels of irrigation; sweet sorghum, a C4 plant that does not need so much water, may play an increasingly important role in the more arid regions of both industrial and developing counties. For any crop species at a particular location, however, the actual productivity in a given year will be strongly influenced by prevailing conditions, including weather, notably rainfall and sunlight.

An annual grass must be planted anew each year; a perennial grass can be harvested while leaving the roots to send up new growth the next year. Roundwood trees like conifers are harvested after they reach their desired

[22] See e.g. Energy Technology Support Unit (ETSU), Department of Energy (now Department of Trade and Industry), *Large Scale Trials of Short Rotation Coppice for Energy, Phase I*, London: ETSU, April 1991, and *Short Rotation Arable Energy Forestry*, London: ETSU, September 1992. ETSU has been engaged for some years, in cooperation with many other interested groups in the UK, in a vigorous research and development programme for energy crops.

[23] Hall et al., 'Biomass for Energy'.

growth, and fresh saplings must be planted to replace them. However, some trees, too, can be harvested just like perennial grasses. Certain species like willow and poplar can be planted simply by inserting fresh bare twigs into the ground, where they take root and grow. Within three to seven years the plant may be six or more metres tall. It can be cut off a few centimetres above root level, and each stump will then produce new growths – several per stump. The process may be repeated several times before the stumps have to be replaced with new planting. This process, called 'coppicing', was a common way to grow firewood in centuries past in many different countries, until it fell out of use as fossil fuels superseded firewood. But coppicing may be on the verge of a renaissance, as research and field trials for so-called 'short-rotation woody coppice' (SRWC) as a source of biomass fuel are burgeoning in Europe and the US. Nor is coppicing the only approach. Whole trees can be grown and harvested by more or less conventional forestry methods, for use either entirely as fuel or for multiple purposes, with the trunks or stems used for timber and the branches and twigs for fuel. Whole trees may be an important option for developing countries, provided environmental considerations can be satisfied. Different crop species will be suitable for biomass uses in different geographical and climatic locations. In particular, species suitable for tropical and sub-tropical regions – including many developing countries – will be different from those suitable for temperate regions, as noted above.

An integrated biomass power station would be centred on what the US DOE calls a 'dedicated feedstock supply system' (DFSS) – that is, an energy plantation: an area of land used to produce a sustainable supply of biomass matched in quality and quantity to the fuel requirement of the power station itself. The prerequisites include choosing one or more species of plant suited to the locality, that will grow rapidly under the prevailing conditions of sunlight, water supply, temperature and soil condition, with a minimum of artificial nutrients and a minimum of artificial control of pests and diseases. The planting regime selected must also take account of local biodiversity, providing for wildlife habitat, including the encouragement of predators that attack pests, and possibly also providing for amenity. Interplanting of different clones of a given species – or possibly even different species – can reduce the hazards of epidemic plant disease or pest infestation associated with large areas of monoculture. Another hazard to be guarded against is fire. Above all, the plantation must be managed according to environmentally acceptable groundrules. One set of such groundrules is proposed in the *Tree Plantation Review* published jointly by Shell International Petroleum Company and the World Wide Fund for Nature in 1993. The final volume, *Guidelines*, includes a section of guidelines on 'The natural environment'; extracts are reproduced here as Box 3.1. In the US the federal government Office of Technology Assessment and the National Audubon Society have published similar guidelines.[24]

3.3 Harvesting and processing biomass

A suitable energy crop regime must also take account of timing, both month by month and year by year, not only for planting and growth but also for harvesting. Some energy crop plants can be harvested with machinery developed for fodder or food crops. Others, like short-rotation coppice, may require modification of forestry machinery, or even machinery designed especially for the purpose.[25] Although a biomass power plant operated on baseload will require a uniform input of fuel, the output of fuel from the energy farms will vary seasonally; the fuelling regime must plan an appropriate schedule for harvesting, to optimize both the growth obtained and the amount of fuel that must be stored for a period lasting at least months. In high latitudes with cold winters, where the ground remains frozen throughout the winter, heavy harvesting machinery can be used for a lengthy period; elsewhere, however, the harvesting period for using such machinery may be much briefer, or lighter machinery may be necessary. In both temperate and tropical regions, logs or coppice stems can be harvested and stored in bundles at the production site, beside an all-weather road, for eventual collection when desired. In such conditions they lose moisture fairly quickly, from say 50% moisture down to

---

[24] *Potential Environmental Impacts of Bioenergy Crop Production*, OTA BP E118, Washington, DC: Office of Technology Assessment, September 1993; *Toward Ecological Guidelines for Large Scale Biomass Energy Development*, New York: National Audubon Society, 1993.

[25] See e.g. *Harvesting Forest Biomass for Energy*, Project Summary 009, London: ETSU Department of Trade and Industry, 1991.

35%, reducing the need for subsequent drying. Herbaceous crops are more difficult to manage, as are wood chippings produced at the production site. Experience of storing such biomass for energy purposes is as yet limited. In its raw state, freshly harvested, it contains so much water that it offers a ready foothold for degenerative pathogens like fungus, which might not only diminish its value as fuel but also constitute a nuisance or even a health hazard. The volumes to be stored may also pose a problem, with adequate inventory management having to be traded off against the cost of storage facilities.

Drying poses further questions. Apart from simple stacking in the open air, one additional possibility would be to provide for simple low-temperature solar drying on the cropland itself. But the cost of even a simple dryer might be disproportionately high compared to the value of the fuel. Alternatively, suitable heat for drying could be made available at the plant itself. The trade-offs must be evaluated, as must the desirability of providing storage facilities at each energy farm as against central storage at the plant itself. In high latitudes, and indeed in some mid-continental areas, transport of biomass from farm to plant in winter conditions may be difficult, at the very time when the plant's electrical output is most in demand. Fuel must be stockpiled at the plant to cover such eventualities, in storage conditions that minimize problems associated with freezing, to which wet biomass fuel will be especially susceptible. In general, the present state of knowledge about drying and storage appears to favour woody crops harvested as bundles and partially air-dried in the open, to be chipped at the power station; but the present state of knowledge is, as noted, far from comprehensive.

Yet another question relates to fuel processing. The fuel feed system into a biomass power plant will require fuel of specified dimensions, usually small – a few centimetres in size – and reasonably uniform, to avoid clogging or plugging of feed lines. Processing biomass fuel to these size specifications is called 'comminution'. Unlike coal, biomass cannot be ground in a mill to pulverize it; it is not brittle enough. Biomass is usually chopped with cutting edges. For some potential energy crop species the technology is already familiar; pulp mills for paper already produce wood chips that are suitable for fuel, although the technology may have to be modified for use with harvested wood from short-rotation coppice, and the drying regime will be different.

Box 3.1 Extracts from the *'Guidelines'* **volume of the Shell/WWF Tree Plantation Review**

Great care should be taken over the precise configuration of the boundary. Unplanted areas within plantations provide much variety and add to wildlife, landscape and amenity value. Those areas which should be excluded from tree planting because of other interests (social, cultural, archaeological, wildlife, etc.) should be identified.

Without compromising the objectives of the plantation, opportunities should be taken to modify its species composition to provide for local needs, by planting or retaining trees or other species of plants which are valued locally or by encouraging animals that may be hunted or fished . . . A plantation inevitably introduces a major new element into the environment and landscape in which it is established. Effects will be highly case-specific and determined by: the type of plantation, management practices, prior status of the site, and risks associated with modifying the edaphic and biotic conditions. Few generalisations are valid for all situations.

There are, in general, two kinds of effects: those relating directly to the productivity and hence success of the plantation enterprise; [and] those that affect other interests either inside or outside the area of the plantation. It is vital for the interests of the plantation manager that those in the former category should be as beneficial as possible. On the other hand, those in the latter will often require a trade-off between intensive management for production and protection of other values.

In order to plan effectively, it is vital to be absolutely clear about the primary and secondary objectives of management. This will assist judgements on, for example, intensity of site preparation and fertiliser inputs and emphasis of labour or capital-intensive methods of silviculture and harvesting. It is recognised that a highly productive, single purpose plantation will have effects, particularly on hydrology and on flora and fauna. Such plantations must be viewed in the full context of regional planning, which is likely to foster a range of such specialised land-uses. In some circumstances, it may be more appropriate to follow a course which is not so highly specialised.

Primary forests tend to protect land against soil erosion and runoff and favour soil processes which enhance fertility. Successful, longterm plantation management should aim to reproduce these beneficial effects. This involves careful assessment of the capability of the site to support repeated harvesting, taking account of nutrient budgets. In contrast to the effects of a plantation on water, climate and biological diversity, the effects on soil condition tend to be confined to the area of the plantation itself. In general, the nutrient status of the soil is best preserved by leaving residue on the site and by disturbing the soil as little as possible, thus reducing compaction, erosion and leaching. Refraining from burning residues helps to reduce losses of nutrients . . . When starting a plantation, it is good practice to ensure that the trees become established as rapidly as possible. This is important both for the productivity of the plantation and to minimise the time during which exposed soil is at risk of erosion . . . Before work begins, sensitive areas should be identified and excluded from planting; examples are wetlands and erodible soils. Consultation with environmental organisations and local interests may be helpful in identifying these. Existing vegetation should be retained where it is advantageous to do so, e.g., as buffer zones to prevent erosion, as filtration strips or as shelter for newly planted trees. Land clearance should be restricted to the minimum necessary for establishment of a productive plantation.

There should be a planned approach to the use of chemicals. Care must be taken to ensure that they are used only when the need is proved, and that advice, controls and regulations are followed scrupulously. There should be consultation with all parties concerned.

If the objective of management is the production of uniform, high quality timber or fibre, the growth of monocultures is appropriate. Otherwise, stands of mixed species are generally to be preferred.

In the planning phase, consultation with other water users and authorities is essential, to determine whether it is necessary to compromise with other interests such as downstream irrigated agriculture. The siting of plantations should conform to national policies for the conservation of biodiversity. Areas of high biological diversity, e.g. existing forests and wetlands, should be avoided. Within the constraints imposed by the need to meet the objectives of the plantation, managers should endeavour to maintain or enhance biodiversity. In deciding the siting of any plantation, consideration should be given to its position relative to external areas of natural forest. The plantation estate may serve as a bridge or corridor for species' movement. Retention of areas of natural forest within the boundaries can facilitate this process, in addition to maintaining biodiversity.

Source: *Shell/WWF Tree Plantation Review*, SIPC/WWF, London, 1993.

Other crops may require different technology for comminution. Some harvesters comminute the crop as it is gathered; others leave the fuel intact for later comminution. As noted above, comminution at the time of harvesting eliminates one processing stage but seriously complicates both drying and storage.

3.4 Electricity from biomass: the options

When the biomass fuel has reached the fuel feed system, attention shifts to the technology of the biomass power plant itself. Once again, the relationship between coal and biomass is surprisingly close. The technologies already used or potentially useful for biomass power generation are also used or being developed for coal-fired power generation, and biomass power may benefit from efforts hitherto focused on coal. Among the technical concepts now being investigated for biomass power generation are:

• direct firing of biomass;
• 'co-firing' of biomass with coal;
• gasification of biomass, for firing gas turbines or diesel engines; and
• pyrolysis of biomass, to produce 'biocrude' liquid fuel to fire diesel engines
  or gas turbines.

As noted earlier, direct firing of biomass is already used in more than 6,000 MW of power stations in the US – some 8,000 MW if urban refuse is included – as well as in many small units elsewhere in the world. The biomass is usually residue fuel of some kind, either from an industrial process on the site – wood pulp manufacture, sugar extraction or the like – or from local urban refuse collection. Units of this type are based on the Rankine steam cycle; the biomass is burned to raise steam for a steam turbine-generator. However, since the supply of residue fuel that can be brought to the plant economically is limited, such units are usually small, perhaps a few megawatts, although pulp mill units may be in the tens of megawatts. So small a steam-cycle unit has a limited efficiency, because in such a small unit the cost of incorporating the reheat and boiler feedwater heater stages necessary for higher efficiency is grossly disproportionate. The steam conditions – temperature and pressure – must be modest, because the more exotic materials required to cope with more severe steam conditions are disproportionately expensive. Accordingly, a small direct-fired biomass power station will have a fuel efficiency of less than 25% – often much less. Only low-cost fuel can be used economically in a unit with such low efficiency. Such low-cost fuel, in the form of limited quantities of residues, can of course be found throughout the world. Forestry residues, for example, are already being used as fuel in about 100 small – 1–2 MW (thermal) – district heating units in Austria, almost all built within the past decade.[26] But direct firing would be more attractive if the conversion technology could be made more efficient. Work to boost the efficiency of direct firing of biomass is already under way, particularly in the US.[27]

An emerging possibility is to 'co-fire' biomass with coal in an existing coal-fired plant. Interest in this option has been stimulated by increasingly tight limits on permitted emissions of sulphur dioxide, substantial quantities of which are produced by coal-fired power stations. One option now being studied is the possibility of mixing a proportion of suitable biomass into the coal feed. Biomass in general contains very little sulphur, less even than so-called 'low-sulphur coal'. Instead of paying a premium price for low-sulphur coal that may have to be transported thousands of miles, a plant operator may be able to obtain a local supply of cheap biomass to blend into the fuel mix, thereby bringing down its average sulphur content. The biomass must either be in a form compatible with the coal-handling equipment on the plant, or be introduced through a separate feed system and burners. Biomass raises different problems of mechanical handling and comminution from those associated with coal, and requires different burners for efficient combustion. A grate-fired boiler may be able to accept both coal and biomass through a common feed system, but a pulverized-fuel boiler will probably require separate systems. The calorific value of the biomass will be lower than that of the coal, reducing the plant's overall output. Nevertheless, these trade-offs may be cheaper than paying a premium price for low-sulphur coal, or backfitting costly flue-gas desulphurization on the plant. As always with biomass, these

---

[26] M. J. Grubb, *Renewable Energy Strategies for Europe*, London: Royal Institute of International Affairs/Earthscan, forthcoming 1994.
[27] US DOE, *Electricity from Biomass*.

questions are acutely site-specific, and must be investigated in each particular context, in respect both of local biomass availability and power-plant characteristics. Preliminary indications are nevertheless that biomass co-firing may prove to be a feasible and attractive way to cut sulphur emissions on at least some existing coal-fired power plants, as the Tennessee Valley Authority in the US has already reported.[28]

Co-firing with coal may also have another role to play. At the moment, building an isolated stand-alone biomass power station runs the risk of creating dependency on a local supply of biomass that is itself unproven; a crop failure could cripple the power station. If, however, the station is built with co-firing capability, it will have access to an alternative fuel supply while the biomass supply system is being proven. Co-firing, to be sure, will entail the extra cost of buying coal and possibly paying some kind of sulphur levy; however, once the biomass supply system has been established and the problems of plant husbandry, harvesting and storage solved, the station can switch over to biomass fuel alone. Co-firing thus diminishes one major risk in building an isolated stand-alone biomass power station, as the US DOE points out.[29] It might be similarly applicable in many different parts of the world.

However, the truly exciting potential for biomass power lies in the direction of fully integrated systems incorporating both fuel supply and conversion plant in a coherent facility designed for the purpose. Consider first the equivalent coal-fired plant. The plant will be designed and built with certain fuel specifications in mind; however, except in the special case of a mine-mouth power station, the supply of fuel need not be considered as an integral feature of the plant. Coal may be brought over long distances if necessary, and many coal suppliers will be keen to compete for contracts to supply the plant, on a long-term, short-term or even spot basis, to appropriate fuel specifications. The same does not apply to a biomass power plant, for two reasons. First, the nature of potential energy crop materials – particularly their low energy density – is such that biomass is likely to be uneconomic to transport over long distances. Second, and partly for that reason, as yet no commercial biomass supply industry equivalent to the coal industry exists. No one will build a

[28] David Tillman et al. 'Co-firing of Biofuels in Coal-fired Boilers: Results of Case Study Analysis', paper presented to EPRI Gasification Conference, San Francisco, October 1993.
[29] US DOE, *Electricity from Biomass*.

biomass power plant without ascertaining beforehand that a suitable supply of biomass will be available for the working lifetime of the plant. Accordingly, for the foreseeable future, anyone planning a biomass power plant must consider not only the conversion technology but also the fuel supply. The implications of this prerequisite will be explored in more detail in later chapters.

Assuming, then, that a site can be found for a power plant in a locality from which an adequate energy crop of biomass can be obtained, what technology will the plant itself employ to generate electricity? Because purpose-grown biomass fuel will cost more than residue fuel, it must be utilized more efficiently, to generate more marketable electricity from each tonne of biomass. The technology options available depend to some extent on the output desired; and that in turn depends on the scale and nature of the available biomass resource. As noted earlier, direct output of heat alone from biomass, for instance for district heating, is already well established over a range of output sizes from hundreds of kilowatts to tens of megawatts. For electricity generation, however, the technical constraints and options are somewhat different. To drive an electricity generator the energy of biomass fuel can be released directly, by combustion, or indirectly, by so-called 'gasification'. If the heat is released directly, it can be used to boil water and operate a Rankine steam cycle turbine-generator, as is done in almost all the biomass power stations already operating. However, the Rankine cycle, as already noted, is very inefficient in the size range under consideration. This size range is determined by the scale of the local biomass resource. In the US, for instance, the DOE proposes units as large as 150 MW; but in more densely populated parts of the world, including Europe and many developing countries, the appropriate sizes may be much smaller, from perhaps 50 MW down to 100 kW or less. For reasons of cycle efficiency and unit capital cost, different output capacities favour different technical options.

One alternative direct-combustion option is to use a so-called Stirling engine to drive the electricity generator. A Stirling engine is an 'external combustion' engine. A gas like helium, or perhaps simply air, is enclosed in a pressure system with pistons. When the gas is heated by burning fuel outside the pressure system, the gas expands against the pistons to turn a shaft. Valves recycle the gas inside the system. The concept was already in use in the late nineteenth century; but it was overtaken by the internal combustion engine

and fell out of use. Nevertheless, research and development continued inter-
mittently, because of the concept's fuel flexibility, low operating emissions
and quiet operation. Recent activities have included efforts to design a Stirling engine to generate electricity by direct combustion of fuels, including
biomass, at sizes from 15 kW to 5 MW. Problems of cost and reliability
remain to be overcome, but the option looks increasingly interesting for small-
scale biomass power applications.[30]

The key to more efficient biomass power generation at the moment, therefore, appears to be not direct combustion of biomass but gasification of
biomass, to produce a fuel gas that can be used in generating technologies
that are efficient and cost-effective at small scales – the diesel engine and the
gas turbine. Biomass cannot, of course, be burned directly in a diesel engine;
nor can it be burned directly in a gas turbine combustor, because the resulting
ash particles, and certain constituents of the combustion gas – notably alkalis
like sodium and potassium – would damage the turbine blades.[31] However,
biomass, like coal, can be 'gasified' – that is, used to produce a combustible
gas as fuel; indeed, biomass can be gasified more easily than coal. The fuel
gas produced can then be burned in a diesel engine or gas turbine, offering
significantly higher efficiency in the appropriate size range, for either electricity generation or cogeneration of electricity and heat, as will be discussed
below.

As already noted, merely heating biomass to a temperature of perhaps 500–
600°C drives off some 80% of the biomass as volatiles – that is, as combustible fuel gas. If the remaining 20% of the biomass reacts with a limited
supply of oxygen, the carbon in the biomass forms not carbon dioxide but
carbon monoxide, which is also combustible. Furthermore, the water in the
biomass reacts to form hydrogen, which is likewise combustible. A mixture
of carbon monoxide and hydrogen is called 'producer gas', 'synthesis gas' or
'syngas'. It was first made from coal in 'town gas' plants almost two centu-

---

[30] See e.g. *The SES Stirling Engine Programme*, London: Sustainable Engine Systems Ltd,
January 1993.
[31] The Allison gas turbines division of General Motors in the US and Daimler Benz in
Germany have research and development programmes for direct combustion of pulverized
biomass in ceramic gas turbines, which are not susceptible to alkali damage; but the technology is in its infancy.

ries ago, and as recently as the 1960s was still distributed in UK cities as a fuel for cooking and heating. It is called 'syngas' because it has long been used by chemical companies as the feedstock for synthesizing other chemicals. The fuel gas from biomass, however, is more complex than coal-based syngas because fuel gas from biomass may contain a significant proportion of complex organic molecules. Fuel gas has been made from biomass in very small, simple gasifiers for a number of years, notably in India and Brazil, for use directly for cooking and lighting. Experience in India has not been entirely satisfactory;[32] that in Brazil – in which producer gas replaces oil in industrial furnaces and boilers – has been better. But interest in modified designs of small-scale gasifiers has recently begun to re-emerge, to produce fuel gas suitable for use in spark-ignition or diesel internal combustion engines, especially for electricity generation for outputs from a few kilowatts up to several megawatts, again notably in India.[33]

Small-scale biomass gasifiers include so-called 'updraft', 'downdraft' and 'cross-current' designs, referring to the direction of airflow through the gasifier, as well as fluidized-bed gasifiers, which are also available in much larger capacities. The updraft design is the simplest. A restricted flow of air enters at the bottom of a chamber filled with suitably sized chunks of wood or other biomass. The biomass at the bottom burns to completion, raising the temperature to several hundred degrees Celsius and heating the biomass above it, driving off volatiles. As the air rises, its oxygen content is depleted, and the upper layer of hot char reacts with the carbon dioxide and steam from the biomass to produce carbon monoxide and hydrogen. The fuel gas emerging from the top of the gasifier contains carbon monoxide, hydrogen and volatiles – hydrocarbons and tar vapour – mixed with nitrogen. The gas can be used directly, for instance for cooking; but feeding it into an internal combustion engine is less satisfactory, because the tars may condense and clog the feed system. A downdraft or cross-current gasifier, although somewhat more complex to design and build than an updraft gasifier, can minimize or eliminate the tar problem. In these designs the air enters from above or from the side, sucked into the gasifier by the lower pressure in the engine feed system. The

[32] H. S. Mukunda et al., 'Open-top Wood Gasifiers', in Johansson et al., *Renewable Energy*.
[33] P. J. Paul and H. S. Mukunda (eds), *Recent Advances in Biomass Gasification and Combustion*, Bangalore: Interline Publishing, 1993.

combustion zone is once again at the bottom of the stack of biomass, and the volatiles are released above this zone; but the airflow, downward or across the bottom, carries the volatiles through the hot combustion zone, cracking the tars into light hydrocarbons that will not condense in the engine feed system. These comparatively simple gasifier designs are attracting increasing attention for generating electricity from biomass, not only in developing countries but even, for instance, in the UK.[34]

A fluidized-bed gasifier has a bed of sand or other inert particles, into which air is blown from below, lifting and 'fluidizing' the particles so that they churn in the airflow like a boiling fluid. To start up, the bed particles are heated to incandescence, perhaps by burning oil or fuel gas in the bed. Once the bed is incandescent any fuel material fed into it – for instance coal or biomass – ignites immediately on contact with the bed; the heat released keeps the bed incandescent, and the startup fuel can be shut off. The concept can be used either for complete combustion or for gasification. For gasification the air inflow is restricted to keep combustion incomplete; gasifying coal usually entails also injecting steam, but the water content of biomass in general supplies enough steam for the gasification reactions. The fuel gas produced is initially tar-laden, like that from an updraft gasifier, and therefore presents a problem for spark-ignition or diesel engines. But the real promise of fluidized-bed gasifiers is more likely to be for use in larger systems, where various approaches to gas cleanup are more feasible and cost-effective, and where the size of the unit will be limited not by the conversion technology but by the available biomass resource.

3.5 Gas turbines and gasification

Fuel gas from biomass, like that from coal, can also be used to fuel a gas turbine. Until the 1980s gas turbines were common in the form of jet engines for aircraft, but in stationary terrestial applications were regarded as a spe-

---

[34] John Seed (Border Biofuels), 'Short Rotation Coppice: A Fuel for the Future', paper presented at Non-food Uses of Land Conference held by National Farmers Union/Friends of the Earth, 20 December 1993. According to Mr Seed, Border Biofuels are applying for permission to build a 5 MW biomass power cogeneration unit that will be based on a downdraft gasifier designed with the involvement of the company.

cialized technology confined to certain special cases, for example as standby generators, or as peaking plant to run only at infrequent intervals of high demand on an electricity system. This is because gas turbines have traditionally used what has been considered premium fuel – either light fuel oil or natural gas. Moreover, a single gas turbine in a conventional configuration – a 'simple cycle' – has historically been an inefficient generator of electricity, because the exhaust gas leaves the gas turbine at a temperature of over 450°C, carrying away a lot of high-quality energy.[35] However, natural gas is no longer a scarce and expensive premium fuel; it is now the fossil fuel for which demand is growing fastest, as new reserves have come on the market in many parts of the world. This cheap and abundant supply has spurred the construction of a lengthening list of baseload electricity generating stations firing natural gas in gas turbines in a more efficient configuration. Instead of a single gas turbine operating alone, one or more gas turbines are linked with a steam turbine. The hot gas turbine exhaust, instead of being wasted, is used in a 'heat recovery steam generator', to boil water and raise steam for the steam turbine. Both gas and steam turbines are coupled to electricity generators. Gas turbine and steam turbine cycles linked together in this way are called 'combined cycles'. Using the most advanced turbine designs now available, a modern natural gas-fired combined cycle station may have an efficiency better than 50%.

However, the world's known reserves of natural gas are mainly in areas already politically volatile and unstable – the Middle East, Russia and North Africa – and other supplies are a long way from their potential markets. Moreover, natural gas prices may need to rise substantially before the necessary pipeline and liquid natural gas (LNG) infrastructure will be put in place. Work has been under way since the early 1980s to develop alternative fuels for combined cycle electricity generation. Thus far the emphasis has been on gasifying coal to produce fuel gas for the gas turbine, in a configuration called 'integrated coal gasification combined cycles', usually abbreviated to IGCC. Several demonstration IGCC plants have already operated successfully since the 1980s, and a series of larger units, ranging in size to over 300

---

[35] The latest designs do better: General Electric, for instance, offers the LM2500 with efficiency of 37% in simple cycle, and the LM6000 with 42% efficiency.

MW (electric), will come on stream in the 1990s. These latest units are based on different designs of large-scale, high-throughput gasifiers, from major industrial firms including Shell, Texaco, Dow, Lurgi, Krupp-Koppers, Tampella and Foster Wheeler.[36]

Most designs of coal gasifier use oxygen rather than air to react with the coal, not so much for technical as for historical reasons unimportant here. Air-blown gasifiers are more appropriate for biomass. An air-blown gasifier produces a fuel gas diluted by the nitrogen of the air. This gas is lower in energy density than that from an oxygen-blown gasifier, and will necessitate appropriate modification of the gas turbine. But the real difficulty arises if sulphur must be removed from such dilute fuel gas, because the sulphur-removal stage must then process large volumes of gas. For coal gasification with an air-blown gasifier this poses a problem, especially sulphur removal at elevated temperature, which is desirable if efficiency penalties are to be minimized. Fuel gas from biomass, however, generally contains very little sulphur. Therefore, assuming the gas turbine can utilize the dilute fuel adequately, biomass may be gasified in an air-blown gasifier, which needs no air-separation unit to produce oxygen. This means that biomass gasification may well be lower in unit capital cost and perhaps somewhat more efficient than coal gasification, especially at smaller plant sizes.

Another influential factor is the operating pressure of the gasifier system. Some gasifier systems work at essentially atmospheric pressure, achieving lower capital cost and simplicity at the expense of lower efficiency. In such a low-pressure system the fuel gas is cooled and cleaned downstream of the gasifier vessel, and then compressed to the pressure needed to feed it into the gas turbine. The gas must be cleaned to avoid carryover of damaging particulates and tars into the compressor; and it must be cooled because technology for cleanup of hot gases is still under development. Biomass power stations based on high-pressure gasifiers are potentially more efficient than stations based on low-pressure systems. High-pressure systems operate with the gasifier pressure sufficiently high to feed fuel gas to the turbine without further compression; a fraction of the high-pressure airflow from the gas

[36] Walter C. Patterson, *Coal-use Technology: New Challenges, New Responses*, London: Financial Times Business Information, 1993.

turbine compressor is diverted through a small booster compressor into the gasifier vessel. This reduces the amount of electricity that the station uses internally, and accordingly raises overall system efficiency. Moreover, a high-pressure system does not need to clean the fuel gas so stringently, provided the temperature of the fuel gas is kept higher than the condensation temperature of the tars and lower than that of alkali vapour.

However, the higher efficiency of high-pressure systems may be offset economically by their greater complexity and higher capital cost. Although hard data are still scarce, both low- and high-pressure systems may find their place in the market, with low-pressure systems dominating the smaller sizes and high-pressure systems becoming more competitive in larger sizes. For reasons of unit capital cost, gas turbine systems smaller than 20 MW may become uncompetitively expensive; and 100 MW represents a pretty large forest. Three designs for high-pressure biomass gasifiers are already in the field, as will be described in Chapter 4. The Finnish engineering firm Ahlstrom originated the Bioflow design, now being developed in an integrated biomass power station in a joint venture with Sydkraft of Sweden. The Institute of Gas Technology (IGT) in the US originated the U-Gas design, now being further developed by Enviropower, a joint venture involving the Finnish engineering firm Tampella and the Swedish electricity utility Vattenfall. An offshoot of the IGT U-Gas design, the Renugas variant, is being demonstrated in Hawaii by a consortium backed by the DOE. A number of companies have demonstrated low-pressure biomass gasifiers, many of which have reliable track records in kraft pulp mills. However, only one company, TPS Termiska Processer, formerly part of Studsvik in Sweden, has demonstrated the gas cleanup technology necessary for a low-pressure biomass power plant. All four designs are now either already in use in pilot facilities for biomass gasification or are in competition for new projects, as will be described in Chapter 4.

3.6 Biocrude

Another route to biomass power is to 'pyrolyze' the biomass – to heat it with very little oxygen present, so that very little oxidation occurs. Under the right conditions of temperature, pressure and other physical factors, the biomass breaks down, producing a high proportion of vapour products that exist in the liquid state at room temperature – so-called 'biocrude'. Like petroleum, biocrude can be refined if desired; but it contains a high percentage of oxygen and requires substantial hydrogenation, and as a source of high-quality road transport fuel is currently economically uncompetitive. It can, however, be burned directly, as fuel either for a slow-speed stationary diesel engine or, if adequate alkali separation proves possible, for a gas turbine. As yet the necessary pyrolysis technology, of several different kinds, is still at the experimental stage; but the biocrude option is also under serious study, particularly in the US.[37] The advantage of taking this liquid fuel route to biomass power is that such liquid fuel can be produced at any time of the year at any suitable location, and stored and transported just like any other hydrocarbon fuel; and its energy density may be high enough to make it worth transporting over longer distances than raw biomass. Pyrolysis introduces a second processing stage between raw biomass and electricity, incurring additional capital costs and process losses; and the environmental impact of burning biocrude will have to be investigated. Nevertheless, producing an intermediate liquid fuel substantially decouples fuel production from power generation; it therefore offers potential for simplification and cost reduction in return for lower overall process efficiency. The liquid fuel approach to biomass power may in due course become more attractive, and is clearly worth further exploration. Although this study focuses on the gasification approach to biomass power, much of its analysis could also apply to the pyrolysis approach. Given the differing characteristics of biocrude for storage, transport, the acceptability of inputs, and so on, the suitability of pyrolysis for power production will depend upon the circumstances: it may ultimately emerge to be the best technology in some circumstances, while gasification or other technologies may remain superior in others. Because pyrolysis technology is in its infancy, saying whether, or when and where, it may prove preferable is as yet impossible. For all these reasons, gasification appears much more immediately promising and is the prime focus of this study.

---

[37] US DOE, *Electricity from Biomass.*

Figure 3.1a Steam-injected gas turbine

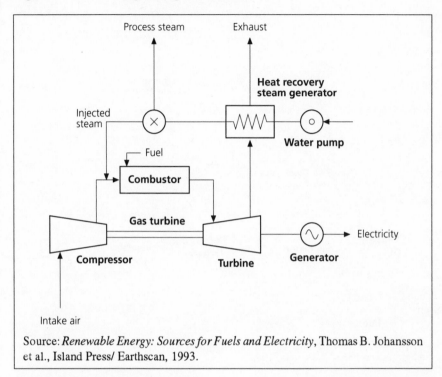

Source: *Renewable Energy: Sources for Fuels and Electricity*, Thomas B. Johansson et al., Island Press/ Earthscan, 1993.

3.7 Advanced gas turbine cycles

A combined cycle plant based on biomass gasification may prove to have an unacceptably high capital cost, partly because the unit cost of the steam cycle will be high in the size range likely to be appropriate for a biomass power plant. Some analysts therefore are looking to gas turbine configurations that can achieve an adequately high efficiency without incorporating a steam turbine.[38] One possibility is to use the gas turbine exhaust to raise steam, and to feed this steam back into the gas turbine itself. The high-quality steam increases the mass flow through the gas turbine expander and boosts its output significantly. This configuration is called a 'steam-injected gas turbine', usually given the acronym STIG; see Figure 3.1a. More than 30 STIG units are

---

[38] Williams and Larson, 'Advanced Gasification-based Biomass Power Generation'.

Figure 3.1b Intercooled steam-injected gas turbine

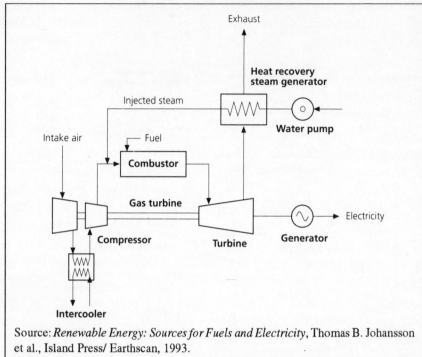

Source: *Renewable Energy: Sources for Fuels and Electricity*, Thomas B. Johansson et al., Island Press/ Earthscan, 1993.

either operating or on order in the US, firing natural gas or distillate fuels. The largest unit commercially available is based on the General Electric LM5000, which produces 33 MW (electric) in simple cycle at 33 per cent efficiency burning natural gas; in STIG configuration the efficiency increases to 40 per cent and the output to 51 MW (electric).[39] This size would be a good match for a biomass gasifier in the size range currently considered to be compatible with fuel supply and transport costs; but the LM5000 cannot yet burn biomass fuel gas with its low energy density. A STIG unit would require a heat-recovery steam generator, but since it would operate at low pressure its cost would not be high. A STIG unit would have a lower overall efficiency than combined cycles, but would also have lower overall unit capital cost; the economics might or might not be better. At the moment no gas turbine could be used in

[39] Ibid.

STIG configuration to swallow all the steam available in a biomass gas turbine plant; the drop in efficiency would therefore be greater than theory suggests.

The key determinant of the thermal efficiency of a gas turbine is the turbine inlet temperature. This in turn is fixed by the metallurgy of the nozzle and turbine blades, and the temperature of the cooling air ducted from the compressor exit into the interior channels of the hot turbine components. One way to boost the output and efficiency of a gas turbine is to add an intercooler – a cooling loop partway along the compressor, to lower the temperature of the partially compressed air. This is the major innovation required for an intercooled steam-injected gas turbine or ISTIG. Reducing the temperature at the inlet of the high-pressure compressor reduces the work the compressor has to do, increases mass flow through the compressor, or both. This bonus itself is largely offset, because more fuel is needed to maintain the firing temperature. However, the lower temperature of the cooling air taken from the compressor means that the turbine inlet temperature can be increased; and this is the key to achieving higher efficiency.[40] The heat recovered from the compressor cooling loop can be fed into the feedwater circuit of a STIG, giving an 'intercooled STIG' or ISTIG; see Figure 3.1b. Although not yet commercially available, ISTIG could be brought to market in three to five years if buyers could be found.[41] Both STIG and ISTIG configurations may have marked advantages for use in biomass gasification power plants of modest capacity, to keep down capital costs while achieving acceptable cycle efficiencies. Neither, however, is yet available in a form suitable for integration with a biomass gasifier. Technological development of biomass power may therefore bypass STIG and move direct from combined cycles to ISTIG.

At its present stage of development, biomass power as a serious energy option in its own right remains to be proven, technically, economically and environmentally. To answer the technical, economic and environmental questions is the role of the first generation of biomass power demonstration plants already in operation or planned, as described in the next chapter.

---

[40] Intercooling is the focus of development work on advanced gas turbine cycles in the US; in particular, EPRI, Pacific Gas and Electric, the US DOE and others have created a Collaborative Advanced Gas Turbine Program whose aim is to demonstrate intercooled combined cycle operation, firing natural gas, before the year 2000.

[41] As footnote 38

Advanced Biomass Power: Projects and Programmes

The concepts for advanced biomass power discussed in preceding chapters have already moved beyond the stage of pure theory. Pilot and demonstration units incorporating biomass gasification to fuel gas turbines are now under development in Scandinavia, the US, Brazil and the EU; indeed, the first such unit, at Vaernamo in Sweden, was commissioned in 1993 and is expected to be fully operational before summer 1994. This chapter will describe the background to the various lines of biomass power development under way, compare and contrast the fuels and technologies involved, and discuss the status of current projects and the programmes of future activity foreseen. The key conversion technologies involved are those based on biomass gasification and gas turbines (see Figure 4.1). Centres of activity include the Renugas process unit in Hawaii, the U-Gas process unit of Tampella in Finland, the Vaernamo project in Sweden based on the Bioflow process, TPS Termiska Processer in Sweden, the Bahia project in northeast Brazil, and Imatran Voima Oy in Finland. Both the US DOE and the European Commission are also becoming actively involved in biomass power demonstration programmes.

4.1 Renugas

The Institute of Gas Technology (IGT), based in Chicago, has been a leading research organization since the 1940s, with the support of major US interests in gas both as fuel and as chemical feedstock. Although historically IGT's interest in gasification focused on gasifying coal, research began in 1977 to develop processes that could produce either fuel gas or chemical synthesis gas specifically from certain forms of biomass. The research started by identifying the relevant differences between coal and biomass as feedstock for gasification. One key criterion was that almost all the biomass carbon, from both volatiles and char, should be converted into gas. Traditional pyrolysis systems like those for making charcoal were not adequate for the purpose.

Figure 4.1 A BIG-GT cycle based on a STIG unit

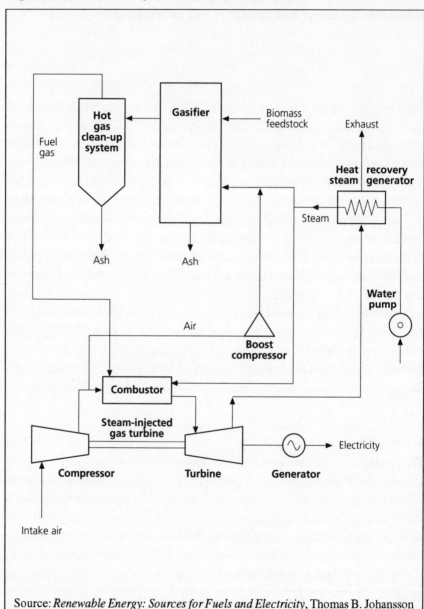

Source: *Renewable Energy: Sources for Fuels and Electricity*, Thomas B. Johansson et al., Island Press/ Earthscan, 1993.

Initial research at IGT sought to determine the process steps that controlled the gasification rate, and to investigate the properties and characteristics of selected biomass feedstocks and their chars. The result was a conceptual process that could gasify biomass in a fluidized bed. IGT patented the process under the name Renugas.[42]

From 1981 to 1987 the DOE supported development of the Renugas process, from laboratory scale upwards, including cold flow models of the critical steps in the process. The data collected were used to design a 'process development unit' (PDU) able to gasify nearly 11 tonnes of biomass per day, at pressures up to 34 atmospheres. The capacity of the PDU was chosen to be one-tenth to one-twentieth that of a commercial gasifier, so that the subsequent scaleup would be reasonable. The PDU operated for two years, carrying out 23 tests and gasifying over 80 tonnes of biomass, including 220 hours of steady-state operation. Short runs were made first, to assess gasification temperature, pressure, feed moisture, feedstock type, steam input, inert bed material, fluidized-bed depth and gas velocity. Longer runs followed, using optimal values, culminating in a three-day continuous run that confirmed the results of the initial tests. The programme also included modifications to the hardware, notably the equipment needed to feed in biomass under pressure. Since the end of the DOE programme the PDU has been used to test various biomass feedstocks for industrial sponsors, and for DOE proposals for a demonstration biomass gasifier plant. Feedstocks tested have included maple wood chips, whole tree chips from two regions of the US, California highway clippings, paddy rice straw, refuse-derived fuel, bark and paper mill sludge and sugar-cane bagasse from Hawaii.

The next stage will be to scale up from the PDU, gasifying 10 tonnes per day, to an engineering development unit (EDU) gasifying 100 tonnes per day. The purpose of the EDU will be to prove the gasification technology itself at commercial scale, and then to couple the gasifier through hot-gas cleanup to a gas turbine-generator, in a complete biomass integrated gasification gas-turbine power plant, usually given the acronym BIG-GT. The DOE, as a component of its National Biomass Power Program (see section 4.6 below),

---

[42] Francis S. Lau et al., 'Development of the IGT Renugas Process', in *Proceedings of the EPRI Conference on Strategic Benefits of Biogas and Waste Fuels*, Palo Alto, California: March 1993.

has joined with the Pacific International Center for High Technology Research (PICHTR) in Hawaii to carry out the scaleup of the Renugas design from PDU to EDU. The EDU, located on the island of Maui at the Paia mill of the Hawaii Commercial and Sugar Company (HC&S), will operate on both wood and bagasse. Other project participants are the IGT, HC&S, the Hawaiian Natural Energy Institute (HNEI), and the Ralph M. Parsons Company, who are the architect–engineers for the project. The first phase of the project – designing, constructing and commissioning the gasifier itself – is already under way, and is scheduled for completion in 1994. The gasifier will be able to use either air or oxygen, at pressures up to 21 atmospheres; feedstocks will include eucalyptus wood, but the plant is to be operated predominantly on sugar-cane bagasse. A later phase will add a 2–3 MW gas turbine and associated gas cleanup technology. The PICHTR project will also include a stage in which the product gas, produced with oxygen and specially treated, will be used not to generate power but to synthesize methanol, itself both a fuel and a chemical feedstock.[43]

4.2 U-Gas

Whereas IGT developed the Renugas process, as its name suggests, explicitly to use renewable biomass fuel, it developed the U-Gas process originally to gasify coal. In 1989 the Finnish engineering company Tampella bought the worldwide licence for the U-Gas process from IGT, and set up its own U-Gas pilot plant in Finland.[44] The 10 MW (thermal) pilot plant, which can operate at up to 30 atmospheres pressure and 1,100°C temperature, was designed for coal, but can also gasify other fuels such as peat and biomass. In 1991–2 the Swedish electrical utility Vattenfall undertook detailed planning and design work for an integrated biomass gasification combined-cycle plant with a thermal input of 140 MW, to be called Project VEGA. Four possible gasifier suppliers were invited to submit pre-studies, and Vattenfall selected Tampella's U-Gas as the most promising. Vattenfall and Tampella formed a joint venture called Enviropower Oy to develop and commercialize this concept.

[43] Ibid.

[44] B. Bodlund et al., 'Power from Biomass in Sweden', in *Proceedings of the EPRI Conference on Strategic Benefits of Biogas and Waste Fuels.*

Enviropower now owns the rights to the U-Gas process and the pilot plant, which has been modified by adding a new feed line and a new product gas cooler, to allow 100% biomass gasification.

The proposed Project VEGA was to have been sited at the Swedish city of Eskilstuna, 100 km west of Stockholm. It would have used 60 tonnes per hour of wet biomass to produce 60 MW (electric) and 60 MW of thermal output for district heating. In June 1992, plans for the Eskilstuna unit were shelved and the project was subsequently abandoned; but Tampella continues to play an active role in biomass gasification development.

4.3 Vaernamo, Sweden

Scandinavia's forest resources also attracted the attention of other major industries in Sweden and Finland, in particular Sydkraft, Sweden's largest privately owned electrical utility, and Ahlstrom, the international engineering firm based in Finland. Sydkraft was seeking to make more effective use of Swedish biomass for power generation; and Ahlstrom had extensive experience of using circulating fluidized-bed (CFB) technology for both combustion and gasification of different fuels, including biomass. Ahlstrom is one of the world's two leading suppliers of circulating fluidized-bed combustion (CFBC) plant, the other being Lurgi – now Lurgi Lentjes Babcock – of Germany.[45] By the mid-1980s Ahlstrom had also supplied four atmospheric pressure gasifiers based on its 'Pyroflow' CFB design, in sizes from 15 to 35 MW (thermal) to pulp and paper industries, for gasifying waste wood; all are still operating, but low oil prices after the mid-1980s eliminated any prospect of further sales.

In 1991 Sydkraft and Ahlstrom joined forces to pursue development of biomass gasification combined-cycle plant for power and cogeneration, using an advanced version of Ahlstrom's CFB gasifier operating under pressure.[46] In September 1991 site work began at the Swedish town of Vaernamo

---

[45] Patterson, *Coal-use Technology*.
[46] 'PCFB Biomass Gasification at Vaernamo', *Modern Power Systems*, January 1993; Ragnar Lundqvist, 'The IGCC Demonstration Plant at Vaernamo', paper presented at seminar on Power Production from Biomass, Laboratory of Fuel and Power Technology of the Technical Research Centre of Finland, Espoo, Finland, 3-4 December 1992.

on what was to become the world's first biomass gasification combined-cycle demonstration plant. The two companies strengthened their relationship in 1992 by forming a joint venture company called Bioflow Ltd, to develop and market the new technology. The Vaernamo plant was commissioned in phases throughout 1993 and will operate at full capacity on biomass fuel by mid-1994. It is designed to operate during the town's heating season, from October to April, supplying 6 MW of electricity and 9 MW of district heating, while serving as a test facility for further development, especially to bring down costs. Because it is the first in the world the Vaernamo plant is conservatively engineered, with some redundancy of subsystems, and of modest size, making its unit capital cost too high to be commercially competitive. Bioflow is confident, nevertheless, that the two-year demonstration phase of work at the plant in 1994–5 will allow them to refine the technology to commerciality.

The Vaernamo plant uses forest and sawmill residues; these are delivered to an external drying plant that uses a separate grate-fired boiler to heat a rotary-drum drier. The drier is oversized for the Vaernamo plant's own requirements because two-thirds of the dried fuel is delivered to other facilities.[47] The crushed biomass is dried from a moisture content of 40–65% down to one of 10–20%, and fed through lockhoppers into the pressurized gasifier. The gasifier operates at a pressure of 22 atmospheres and a temperature of 950–1000°C, to yield a product gas with an energy content about one-eighth of that of natural gas. The fuel gas passes through a cyclone that strips most of the solids and returns them to the bottom of the gasifier. The operating temperature in the gasifier is high enough to crack heavier tars, so that they do not condense when the fuel gas is then cooled to 350°C to pass through a ceramic filter to remove residual particulates. Dolomite in the bed material of the system also provides some catalytic cracking to augment the thermal cracking.

The clean fuel gas is then burned in the gas-turbine combustor to generate 4.1 MW of electricity. The gas-turbine compressor also pressurizes the gasifier. The hot exhaust gas from the gas turbine enters a heat-recovery steam generator to produce steam for the steam turbine, which generates a further 2 MW

---

[47] A commercial design could have an integrated drier using heat from the plant itself: either flue gas from the heat recovery steam generator or low pressure steam from the steam cycle.

of electricity. Heat for district heating is supplied from the steam-turbine condenser as well as from other cooling circuits in the plant.

Bioflow is now leading one of the two design teams competing for the Bahia project in Brazil (see section 4.5 below).

4.4 TPS Termiska Processer

TPS Termiska Processer is an independent Swedish company, formerly part of the Swedish energy engineering group Studsvik and now owned by a group of Swedish municipal energy suppliers, biomass energy interests and the Federation of Swedish Farmers. For more than ten years the group, now known as TPS, has been involved in advanced combustion and gasification technology, with biomass including wood fuels and refuse-derived fuels (RDF) figuring prominently. TPS developed a design for a circulating fluidized-bed gasifier operating at atmospheric pressure, suitable for gasifying biomass including RDF; two 15 MW (thermal) units based on the design have been operating at Greve-in-Chianti in Italy since 1991, gasifying RDF. TPS has placed particular emphasis on gas cleanup technology, including a proprietary tar-cracking technology essential if the fuel gas from the atmospheric-pressure gasifier is to be compressed to feed to a gas turbine. A TPS design formed the basis of an budget proposal submitted to Vattenfall by Kvaerner Generator for the VEGA project described in section 4.2 above. TPS also did a design study for Gullspang Kraft AB of a BIG-GT unit using the TPS gasifier. TPS is leading the design team competing with Bioflow for the Bahia project in Brazil (see section 4.5 below).

4.5 Bahia, Brazil

Brazil is a leading producer of renewable energy: more than 90% of its electricity is hydroelectric, and almost a third of its total primary energy supply comes from biomass. In the northeastern region of Brazil, electricity is supplied by Companhia Hidro Eletrica do São Francisco (CHESF), a subsidiary of the national utility Eletrobras. Electricity use in the region is rising rapidly; but the low-cost hydro resources of the region will be fully utilized by the turn of the century. Further expansion of hydroelectricity would entail installations in the environmentally sensitive region of the Amazon River farther north, involving both major controversy and significant cost increases. Beginning in the early 1980s, therefore, CHESF has carried out detailed studies of the potential for generating electricity from biomass in northeastern Brazil, initially from sugar-cane bagasse and other residues, and then additionally from plantation wood. The studies were drawn together in an impressively thorough report initially published in July 1992.[48]

The potential appears to be substantial; but the necessary practical experience, both for producing the fuel and for converting it to electricity, still remains to be gathered. The creation by the UN of the GEF, administered by the World Bank, has provided the impetus to take the concept farther. The mandate of the GEF is to promote investment in four key areas of global environmental importance: protection of the ozone layer; support for biological diversity; maintenance of international bodies of water; and control of emissions of carbon dioxide to the atmosphere. Although the GEF has only limited scope to intervene in the fossil energy sector, it can assist in accelerating the development of those renewable energy technologies judged to be sufficiently close to commercialization. Biomass integrated gasification gas-turbine (BIG-GT) technology could make a significant impact on the carbon cycle by replacing fossil-fuelled electricity generation with biomass-fuelled electricity generation with no net emissions of carbon dioxide to the atmosphere. Accordingly, the GEF has made substantial funds available to accelerate development of BIG-GT, and northeastern Brazil has become a focus of such activity.

In June 1991 the then Secretary of State for Science and Technology in Brazil invited 18 local Brazilian and multinational organizations to a meeting to create a project development group. Five organizations expressed an interest in further studies: CHESF, Eletrobras, Companhia Vale do Rio Doce (CVRD), Fundacao de Ciencia e Tecnologia (CIENTEC) and Shell Brasil. Phase I, a preliminary investigation, was completed in March 1992. On the basis of the intermediate and final reports submitted in Phase I, the GEF confirmed that it would make available two tranches of grant funding: US$7.7 million for two years of process development in Phase II, and US$23 million

---

[48] A.E. Carpentieri et al., 'Future Biomass-based Electricity Supply in Northeast Brazil', *Biomass and Bioenergy*, Vol. 4, No. 3, 1993, pp. 149–73.

for implementation of the plan in Phase III, which would entail constructing and operating a 30 MW (electric) demonstration BIG-GT unit in Bahia.[49]

At the time not enough was known to define the optimum configuration for BIG-GT technology, including gasifier pressure, gas cleanup and gas turbine cycle. Since the available technical and economic data were too 'soft' to permit competitive tendering between options, Phase II of the project includes two parallel and competing lines of development. Two independent project teams are working on distinct technology packages. Each team is led by a gasifier system developer: TPS Termiska Processer for the low-pressure gasification option, and Bioflow for the high-pressure gasification option. The question still at issue is whether the efficiency advantage of high pressure over low pressure will offset the additional problems of reliability and specific investment cost expected with high-pressure systems at 30 MW capacity, especially given the constraints of time and funding for Phase II. Both teams are working with General Electric, which is adapting its LM2500 gas turbine for use in BIG-GT. Phase II will end in 1994 with the choice of one of the two systems for implementation in Phase III.

The successful project team chosen in 1994 will complete the full basic engineering design. On the present schedule, before the end of 1994 a comprehensive information package will be available, together with a project evaluation document, as a basis for Phase III – actual construction of the Bahia plant. The original five participants in Phases II and III have options for equity roles and are expected to dominate the joint-venture company that will be formed to build, own and operate the plant. Third-party equity participants may also have a role in the joint venture; these may include, for example, other power utilities and independent generators, both in Brazil and elsewhere, to get access to operating experience; portfolio investors; biomass producers such as the Brazilian sugar-cane industry, which is looking for better options for utilizing biomass; equipment manufacturers including the gasifier suppliers; and carbon dioxide producers interested in offsetting carbon dioxide production elsewhere. The coming stages of the Bahia project will be closely watched, both in Brazil and around the world.

---

[49] Elliott and Booth, *Brazilian Biomass Power Demonstration Project.*

4.6 US National Biomass Power Program

According to the USDOE:

> More than any other energy technology, biomass power is capable of contributing to the nation's energy needs while decoupling energy production from environmental degradation. Expanded investment in biomass energy technology will substantially impact the US economy by creating new income and jobs, strengthening US industrial competitiveness, and revitalizing economically depressed areas in rural America. These economic benefits will be realized through an environmentally sound renewable energy technology.[50]

> The US DOE's National Biomass Power Program, launched in 1991, is founded on a collaborative strategy involving industry, the research and development community, regulators, potential users, and state and federal agencies. The Program goal is to add 6,000 MW (electric) of commercially competitive renewable biomass power capacity to the US power generating base by the year 2000. With the help of advanced biomass power technologies and new dedicated feedstock supply systems (DFSSs), as much as 50,000 MW (electric) of biomass power could be in place by the year 2010.[51]

The immediate objective of the US National Biomass Power Program, to have 6,000 MW of new biomass power in operation by the year 2000, is certainly ambitious; the longer-term vision for 2010 – the DOE projects 25,000 MW and the US Electric Power Research Institute (EPRI) as much as 50,000 MW – frankly breathtaking. As yet, however, its practical fruits have barely begun to materialize; the five-year plan laid out in the document from which the above quotations are taken is proposed to run from 1994 to 1998. In the spring of 1993, EPRI, the research arm of the US investor-owned electricity supply industry, joined forces with the DOE to offer to share the costs of innovative biomass power plant projects, initially as case studies of 'sustainable biomass plantations/power plant systems'. The case studies 'will evaluate options, sites, costs, and risks, and will develop system design/cost, en-

---

[50] US DOE, *Electricity from Biomass.*
[51] Ibid.

vironmental risk management plan, and schedule for a biomass electric power supply system. If studies suggest that such a system would be a prudent investment for the host utility, the utility is expected to submit a proposal to EPRI and the DOE for minor cost share in actual establishment of a crop and construction of a power plant.'[52] By late 1993 some 25 proposals had been received; indications were that about ten would be supported. EPRI and DOE added that case studies would be selected for co-funding using the following criteria:

• intent and capability of utility to pursue project if case study is favourable;
• involvement of farmers, equipment vendors, manufacturers, power plant designers and builders, state resource agencies, etc.;
• availability of suitable cropland;
• economic considerations such as costs, markets;
• market size: crop diversity, spectrum of opportunities and business alliances, total MWe potential, other factors.[53]

Meanwhile the work of the DOE and the National Renewable Energy Laboratory on the National Biomass Power Program continues.[54]

4.7 European Community (now European Union): The THERMIE Programme

In 1990 the Directorate-General for Energy (DGXVII) of the European Commission launched a programme called THERMIE, with a budget of 700 million Ecu from 1990 to 1994, to foster innovative non-nuclear energy technologies, under four main headings: rational use of energy, hydrocarbons, solid fuels and renewable energy. Unlike the JOULE programmes of the Directorate-General for Research and Development (DGXII), the THERMIE programme was prepared, at least in principle, to offer financial support to projects involving major investment, up to the scale of demonstration plants. One of the projects eventually granted THERMIE support was the 300 MW Puertollano integrated coal-gasification combined-cycle (IGCC) plant, now

---

[52] *Host Utility*, Electric Power Research Institute, Palo Alto, California: March 1993
[53] Ibid.
[54] Ralph P. Overend and Richard L. Bain, 'The DOE/NREL National Biomass Power Program: Gasification Project Updates', paper presented to EPRI Gasification Conference, San Francisco, October 1993.

under construction in Spain, at an estimated cost of some US$860 million. THERMIE contributed an initial tranche of 30 million Ecu, followed by a further 15 million Ecu; an additional 15 million Ecu is now being considered.

In 1993 THERMIE programme administrators let it be known that they were open to bids for support for biomass gasification projects. As this is written no formal proposals have thus far been announced; but biomass advocates in the UK, including the Energy Technology Support Unit (ETSU) of the Department of Trade and Industry, are known to be keenly interested in promoting a biomass gasification project in the UK, and several other member states, notably Denmark, will almost certainly show similar interest. Within the EU, gasifier suppliers offering technology potentially suitable for biomass gasification include the Coal Research Establishment of British Coal, the UK–German partnership of British Gas–Lurgi with its 'slagging gasifier', Lurgi itself with various fluidized-bed gasifier designs, and the German partnership of Rheinbraun–Uhde with its 'high-temperature Winkler' design, which has already been used to gasify peat in Finland. However, the first demonstration unit may well select a gasifier from outside the EU, perhaps the TPS or Bioflow design, or may elect to use one of the simpler downdraft or cross-current designs. At the same time, member states including the UK, Denmark and Italy are pursuing field trials of short-rotation coppicing and other potential energy crops suitable for biomass power;[55] and interest is gaining momentum. It appears to be particularly vigorous in the UK, where the government's 'Non-Fossil Fuel Obligation' (NFFO) arrangement has invited bids specifically for biomass energy based on short-rotation coppice, to receive NFFO support.

Several member states have expressed an interest, either through government agencies or through non-governmental bodies, in more advanced biomass power systems; but as yet the concept remains essentially theoretical, even in Denmark, the EU member with probably the strongest national commitment to biomass power.[56] The scale and focus of these activities differ markedly

[55] See e.g. the newsletter *Wood Fuel Now* published by ETSU, and the Project Summaries published by the Renewable Energy Enquiries Bureau of ETSU.
[56] Danish utilities ELSAM and ELKRAFT have built a series of 15 steam cycle power and cogeneration stations burning biomass, notably including two that burn compressed straw from Danish farms; see *Large-scale Bioenergy: Challenges and Risks*, ELSAM, Copenhagen: 1993.

from one state to another, but everywhere projects pertaining to biomass as a solid fuel are small-scale (at most a few megawatts): district heating or combined heat and power (CHP) using biomass residues in conventional steam-raising boilers.[57] Some relevant research and development towards advanced systems is however taking place.[58]

Two prospective members of the EU, Finland and Sweden, also have national biomass programmes including various approaches to biomass power. One distinctive concept is that being pursued by the Finnish utility Imatran Voima Oy (IVO). IVO notes that congeneration applications for biomass power, in Finland and similar candidate countries, may be limited by the size of the available heat load; in Finland and some other countries existing CHP facilities already account for almost all this load. IVO has sought to develop a gasification/gas-turbine cogeneration design whose power-to-heat output ratio delivers a higher proportion of electricity, particularly for retrofitting existing units. The IVO concept includes a high-pressure drying unit integrated with the power-generation cycles. Biomass is dried by steam at high pressure, and the additional high-pressure steam produced from the biomass is fed into the steam cycle, either in a suitable STIG configuration or in combined cycles. The concept is called IVOSDIG, for 'IVO steam-dried integrated gasification'. It can be used for any moist fuel, including not only biomass but also peat and lignite. Tests on peat and biomass have been under way since 1991, and IVO has conducted pre-feasibility studies for an IVOSDIG retrofit on an existing peat-fired unit in Finland.[59] There as elsewhere, however, the more advanced concepts for biomass power are still in the development phase.

---

[57] *Energy and Biomass: Project Paper No. 2. An Overview of Activities in Member States and Prospective Member States*, Strasbourg, Scientific and Technical Options Assessment (STOA), European Parliament, August 1993.

[58] See e.g. W. M. Dawson, *A Small-scale Gasifier-based Combined Heat and Power System*, Northern Ireland Horticultural and Plant Breeding Station, Loughgall, Northern Ireland: 1993, describing a unit using a downdraft gasifier coupled to a diesel engine.

[59] S. Hulkkonen et al., 'Development of an Advanced Gasification Process for Moist Fuels', paper presented to EPRI Gasification Conference, San Francisco: October 1993.

Designing an Integrated Biomass Power Station

5.1 The fuel

Planning an advanced biomass power station anywhere in the world will entail considering an unusually wide range of factors – some familiar to electricity planners, others much less so. The first requirement is a site; and the site in turn will determine the range of possible biomass fuels. Of course, if the biomass power unit is to use residues alone, the choice of fuel is predetermined. If the unit is planned as a system integrated with another biomass application, such as pulp production or co-production of alcohol, the relevant crop will influence the design of the corresponding biomass power unit. However, for a free-standing biomass power station fuelled with an energy crop, the first crucial decision will be the choice of crop plant or plants. No matter who is responsible for preparing the land and growing and harvesting the energy crop, factors relating to the site itself – the climate, sunlight, temperatures, water supply and soil condition – will define the range of plant species and varieties that may be suitable, as already indicated. Temperate or tropical climate, wet or arid conditions, good-quality or degraded soil – these major factors will narrow and focus the choice of energy crop suitable for a particular site. Even so, success is not guaranteed. Even in industrial countries with long histories of agriculture for food production, prospective biomass fuel growers can call on comparatively little experience of growing energy crops. In developing countries certain well-known crops like sugar cane may also become important energy crops for biomass power, but experience with other possible energy crops is still limited. Only extensive practical experience of actual crops in the actual growing conditions at a particular site can give reliable data for the performance and productivity of particular species and varieties. That experience is accumulating rapidly; but caution is still necessary.

Planners of a biomass power station must come to suitable arrangements for fuel supply, which will differ in key respects from those for a coal-fired station. As indicated earlier, the size of a biomass power station will be determined to a large extent by the biomass fuel resource that can be obtained locally at acceptable cost, including transport cost. Various arrangements are possible. The station owners may themselves own the land and grow the energy crops to fuel the station. However, experience in building and operating power stations does not usually encompass experience of agriculture or forestry; the station owners may wish to establish joint ventures or consortia that include experienced farming or forestry groups. Given the importance of environmental considerations and constraints for biomass activities, such consortia should also involve environmental specialists. Alternatively, the station owners may negotiate appropriate long-term contracts with local landowners, farmers and agricultural merchants, entering commitments to buy a stipulated amount of energy-crop fuel of stipulated quality and characteristics, on a stipulated timetable at stipulated prices over the working life of the plant. Such long-term biomass fuel supply contracts will have certain attributes in common with long-term coal supply contracts; but, unlike coal suppliers, biomass fuel suppliers may not have the option of buying in fuel from suppliers elsewhere if necessary to fulfil their contractual obligations, and adequate legal and operating provisions will therefore have to be made for failure to meet supply obligations. Unlike coal, energy crops may suffer from pests, diseases or fire; the station owners may have to arrange take-or-pay contracts to provide a surplus of fuel above station peak demand to cover such contingencies. Herbaceous crops must be handled differently from woody crops; the station owners will have to have a policy on fuel inventory, taking account also of factors like bulk density.

In the US, where land is available in comparatively large tracts, and a biomass power station could be constructed on a scale that would not be feasible in Europe or in most other parts of the world, the DOE offers a striking illustration of the land requirement for a biomass power plant with an output of 150 MW.[60] According to the DOE, one hundred 'energy farms', each a plantation of one square mile, sited within a 25-mile radius of the power plant, could fuel the plant sustainably, while covering only 5% of the land area within this radius around the power station. In Europe, with its higher population density and more intensive settlement patterns, even the largest biomass power

[60] US DOE, *Electricity from Biomass*.

stations would be smaller than this, perhaps a maximum of 100 MW, and most would be smaller still; the radius of a comparable circle of energy crop plantations would therefore be likewise appropriately smaller. The proportion of land necessary would be no larger, and transport costs lower. In developing countries yet smaller stations might be more appropriate, right down to one megawatt or less, especially in rural areas, with land use requirements to match. One important proviso must, however, be noted. The DOE estimate for land use may be unduly optimistic about the biomass productivity achievable in the cluster of energy farms around its 150 MW unit. Measured yields for particular species at a given location can vary widely, depending on weather, rainfall, soil water retention and other variables. If the requisite productivity is not achieved, either more land will be required or the station's output will be less than its design output, or both, to the detriment of its economic status. Similar considerations will apply to any biomass power station anywhere in the world, no matter what its size or its fuel.

Producing coal, however responsibly, inevitably raises issues of environmental impact, sometimes severe; these may include water or air pollution, noise, subsidence, and land degradation. Producing biomass, too, raises issues of environmental impact, some shared with coal and others distinctively associated with growing, harvesting and processing plants for fuel. Whereas extracting coal is a once-and-for-all operation, extracting biomass can and should be done 'sustainably', so that it can be continued indefinitely if desired, without degrading the local environment. Managing biomass sustainably entails:

• minimizing the use of artificial nutrients or pesticides, to avoid polluting local waterways or groundwater, or damaging local ecosystems;
• maintaining soil condition, including tilth and natural nutrients, possibly by returning some biomass to the fields;
• organizing planting to avoid or minimize the problems of monoculture for wildlife habitat, pests and diseases; and
• organizing harvesting to minimize impacts on wildlife and to improve visual appeal.

The environmental impact of biomass production and use in both industrial and developing countries has been under investigation for some years in many parts of the world, and the literature is already extensive. As already noted in Chapter 3, Shell and the World Wide Fund for Nature, the US government Office of Technology Assessment and the National Audubon Society have published guidelines for biomass development that include detailed criteria for environmental management in any part of the world. In the UK the Energy Technology Support Unit of the Department of Trade and Industry has produced a series of studies on environmental impacts and environmental management.[61] Nevertheless, environmental standards and criteria for biomass activities continue to evolve rapidly. Apart from the impacts of cultivating, harvesting and utilizing biomass fuel, an additional environmental impact that may be particularly significant in industrial countries is that of the transport movements required to deliver biomass fuel from plantation to station. The United Nations Environment Programme notes that biomass use in developing countries also raises concerns about deforestation, soil erosion and desertification, as well as questions about the conflict between producing food and producing fuel, and offers guidelines to maximize opportunities while minimizing undesirable side-effects.[62] Scrupulous attention to environmental considerations will be important not only for the sake of the environment itself but also for the long-run acceptability and success of the technology.

Some important environmental impacts of biomass activities may be not negative but positive. Crops grown for combustion can be interplanted and harvested together; unlike food crops, fuel crops may not necessarily involve monoculture at all, and will almost certainly involve less monoculture than current agricultural patterns for food production in industrial countries, which can cause serious damage to soil structure. The edges of energy crop plantations may offer better habitat for wildlife than those around fields producing food crops. In the context of fuel crops, the concept of sustainability is still under investigation.[63] Thus far, nevertheless, it appears to be feasible, sharing the attributes of other sustainable regimes of cultivation. To date, efforts to improve the genetic stock of potential energy-crop plants have been lim-

---

[61] See e.g. G. E. Richards (ed.), *Wood: Energy and the Environment*, London: ETSU/DTI, September 1992.
[62] *Green Energy: Biomass Fuels and the Environment*, New York: United Nations Environment Programme, 1991.
[63] Hall et al., 'Biomass for Energy'.

ited,[64] but results on energy-crop plants including willow, poplar and sugar cane have been encouraging, and efforts are accelerating in many parts of the world, on a widening range of species. If biomass power develops as many anticipate, crossbreeding, selection of better clones and even genetic engineering will widen and enhance the range of energy-crop plants available, offering higher biomass productivity, lower water and nutrient requirements, better resistance to pests and diseases, and other advantages.

5.2 The generating technology

Like a coal-fired power station a biomass power station must have adequate land area with suitable foundations for buildings including those necessary for fuel handling and storage, the generating plant itself and a substation to connect it to the transmission system, with planning permission for construction and operation. Like a coal-fired station it must have adequate transport connections for delivery of fuel, by rail, road or possibly barge: for instance, a station using 200,000 tonnes of raw biomass a year will need some 660 tonnes – perhaps 30 truckloads – a day. The station must have access to the local electricity grid; and it may need a water supply for cooling water. Unlike a coal-fired station, however, a biomass station must have some form of guaranteed fuel supply for its operating life, since unlike coal no general market exists for biomass fuel in the quantities the station will require. The fuel supply may be drawn from available residues, if the residues arising are sufficient to fuel the plant; but a more likely source will be dedicated energy crops from the surrounding area. One possibility will be to arrange to fuel the plant with residues while energy crops like short-rotation coppice or swift-growing grasses are being established, with the intention of blending in or completely switching to energy-crop fuel when it becomes available. Such fuel flexibility, however, will necessitate including fuel-handling equipment that may be more expensive, able to cope with both residue fuel and energy-crop fuel of whatever kind is chosen. Planners of advanced coal-fired plant, for instance CFBC plant, already have to weigh the advantages and disadvantages of building in fuel flexibility, and the anticipated trade-offs between

---

[64] Ibid.

more expensive equipment and cheaper fuel; planners of biomass plant can learn from coal-firing experience in this respect. Yet another option, now being promoted in the US, is to develop co-firing projects for existing coal-fired power stations. Energy crops can be planted near a coal-fired power station and used for co-firing in the coal-fired station, until the crop regime – species, husbandry, harvesting, delivery – is fully established. Then the biomass fuel can be switched into a dedicated biomass power station built next to the coal-fired station. Common infrastructure shared by the two stations would reduce capital costs.

Even in the case of a coal-fired station, the fuel must meet certain specifications arising from the generating technology. In the case of a biomass power station, the fuel specifications may be yet more stringent, depending on the conversion technology to be used. As mentioned earlier, direct firing of biomass is relatively simple, but is inefficient, and likely to be uneconomic except with low-cost residue fuel or possibly in the context of a co-firing project. More efficient conversion technology may be more complex, and may entail tighter fuel specifications. Moreover, although simple direct burning uses well-established technology, more advanced and efficient conversion technology, for instance a high-throughput efficient gasifier coupled directly to a gas turbine in combined-cycle, STIG or ISTIG configuration, is not as yet fully demonstrated even at pilot-plant scale. Accordingly, the next phase of biomass power development must encompass one or more demonstration plants incorporating the latest gasifier and gas turbine technology, matched to the necessary biomass fuel. As outlined in the preceding chapter, this phase is already under way in several different countries, with participation from major companies, governments and international agencies.

Although biomass has some advantages over coal with respect to waste products – biomass, for instance, contains little sulphur or heavy metals – a biomass power station will still produce gaseous, liquid and solid wastes. Managing these wastes should not be difficult. Depending on the conversion technology employed, a biomass power station can install standard pollution-control equipment like that used to clean up gaseous and liquid effluents from conventional coal-fired stations or the new advanced 'clean coal technology' stations, benefiting from the experience already accumulated. Depending to some extent on the particular biomass fuel and conversion technology, the solid waste from biomass is likely to be at least benign, and may indeed contain nutrients that can be beneficially returned to the land from which the biomass was harvested. The only significant gaseous pollutant is likely to be nitrogen oxides, usually called NO_x. NO_x arising from nitrogen in fuel should not be a problem in low-pressure systems; catalytic removal of fuel NO_x in high-pressure systems may need further research and development. NO_x from nitrogen in the air is always a factor in gas-turbine systems, because the comparatively high flame temperature in a gas-turbine combustor oxidizes the nitrogen in the air into so-called 'thermal NO_x'. However, the flame temperature from low energy-density fuel gas from biomass is similar to that in new designs of low-NO_x burners for gas-turbines, so thermal NO_x levels should be relatively low.

Project planning for a biomass power station must also examine the impact not only of the station but of its fuel supply on local employment and the local economy. Maintaining the fuel supply may create more jobs than operating the power station. Compared with a traditional coal-fired station, a biomass power station of any design will be small – from perhaps a few megawatts up to a maximum of 150 MW in some parts of the US, probably less than 100 MW in other industrial countries or developing countries – to keep fuel transport costs low enough. From a traditional viewpoint that might be considered a disadvantage; but in fact a strong trend has developed in many countries away from gigantic power stations towards smaller units. A smaller station is easier to site and makes more tolerable demands on land, water and amenity. With technologies now available it can be much cleaner and more environmentally acceptable; it may even be sited comparatively close to users, cutting transmission and distribution costs and losses, and making cogeneration or district heating easier. Its output can be planned within a plausible forecasting horizon. It can be brought on-stream just when it is needed, reducing the drain on the system from costly unused surplus generating capacity; and it can rapidly generate cash flow from saleable electricity. A small station can use technology that is modular and can be fabricated mostly in factories rather than under site conditions, reducing the problems of quality control and labour relations that caused expensive delays and overruns on many stations in the 1960s and 1970s. A modular design makes maintenance simpler, and can be replicated as desired, to increase capacity in appropriate steps. The concept of biomass power follows this trend. Biomass power will lend itself in due course to a widespread network of small stations,[65] decentralized throughout rural agricultural, forestry or perhaps especially marginal areas, bringing a whole new modern economic activity of tolerable scale into such areas, with extra jobs and income, helping to counter the trend of population movement away from the land into the cities. If biomass power can establish itself as a credible commercial and economic concept, similar considerations may come to apply not only to rural areas of industrial countries but possibly – and even more importantly – also to rural areas of developing countries.

5.3 Afterthought or forethought?

Biomass power at its most basic – burning residue fuels in conventional steam plant – could certainly continue to expand, in an essentially ad hoc manner, wherever suitable residues are reliably available at a low enough cost. But a truly significant role for biomass as a modern fuel for electricity generation will depend on a much more coherent forward view, in which all the requisites – dedicated energy crops, biomass fuel processing, and efficient modern conversion technology generating electricity at a commercially competitive cost – come together in a complete system. Advanced biomass power looks good on paper, as many commentators have shown. Nevertheless, to establish it as a realistic and practical energy option for the coming century will still require substantial commitments of effort, money and time. The next chapter will discuss how such commitments might be marshalled and applied.

---

[65] Probably also including other renewable technologies like windfarms.

The Learning Curve

6.1 Development strategies

Thus far, work on advanced biomass power has aimed to prove the necessary technical concepts: processing biomass as feedstock for a gasifier; feeding various forms of processed biomass into various types of gasifier; gasifying different feedstocks in different gasifiers; and most recently firing gas turbines with fuel gas from biomass. The projects and programmes already under way or soon to materialize will gather performance data, and endeavour to rectify the inevitable technical problems that prototype and demonstration plants reveal. However, even assuming that the basic concepts can be shown to be technically feasible, the longer-term challenge will be to develop systems that not only work but are economically competitive. To do so will require development strategies. Factors that affect the economic status of biomass power will differ from place to place; they include the present and projected uses of electricity, existing electricity supply systems, governmental and financial institutions, economic activities, agricultural and other subsidies, and the prices and availabilities of competing electricity supply technologies and fuels. An effective development strategy for biomass power in any part of the world must take account of these local factors. Effective strategies for, say, the US, the UK, the EU, Brazil, India or China would differ markedly. To give but one example: in many industrial countries, the most serious competition for biomass power in the foreseeable future will come from power stations burning natural gas. In many developing countries, however, such competition does not exist, nor is it likely to arise in the near future. Development strategies for biomass power must be shaped accordingly.

In any region of the world, merely demonstrating that biomass can be grown sustainably, suitably processed, gasified and fired in a diesel engine or gas turbine to generate electricity will not suffice. A mature biomass power industry will emerge only if the concept of biomass power can bring together a remarkably disparate range of interests, and compete successfully against other options for generating electricity, probably in many different geographical locations each with its particular local characteristics. What might such a mature biomass power industry look like? It would probably be based on a dispersed infrastructure of fuel suppliers growing energy crops over a wide area, with biomass power stations – some perhaps quite small, one megawatt or less, others larger – likewise distributed throughout the area. In industrial countries the power stations would probably be connected to the electricity grid, and in temperate latitudes might also supply heat for district heating. In the rural areas of developing countries such as India, China or Brazil, on the other hand, biomass power stations might be stand-alone units serving individual villages or local clusters of villages. At one extreme of decentralization, individual fuel suppliers might be private farmers or foresters, selling their raw biomass fuel to the highest nearby bidder among the power station operators, on a short-term basis. At the other extreme, fuel suppliers might be equity participants in power stations, along with electricity supply companies and possibly also equipment manufacturers, as integrated consortia embracing the entire system. Between these extremes many different permutations can be envisaged, with different legal and contractual relations between the main players. To establish such a mature and widespread industry should be the aim of a biomass power development strategy, wherever it may be pursued.

Who might be involved in devising and implementing an appropriate strategy? Recall the roster of potential beneficiaries listed in Chapter 2 above (section 2.3). Some of these groups will look on biomass power as a source of income; some as a source of jobs; some as a source of regional or environmental benefits; some as a source of energy security or enhanced international trade or foreign exchange savings. At the moment, however, very few are aware of biomass power at all. The first stage in any strategy must therefore be to disseminate information and build a constituency – to make these potential beneficiaries aware of biomass power as an option, aware of its current status and prospects, to engage their interest and involve them as active participants in further developments.

Those already aware of biomass power are also aware of the risks it now entails. Those who might grow energy crops do not know if their produce will find a market at an acceptable price. Those who might build biomass power stations do not know if they can buy suitable fuel at an acceptable price; they do not know if the technologies available will be sufficiently cheap and reliable; and they do not know if they can sell the electricity at a price high enough to recover their costs and make a profit. Moreover, such uncertainties look different in different places. The potential and the problems in, say, Brazil, are different from those in the UK, and the same holds true across the world. As noted earlier, the dependence of biomass power on a fuel supply from the immediate locality makes it more site-specific than fossil-fuel options. Each locality has to be considered on its own terms – climatic, geographical, social and economic. Nevertheless, successful resolution of uncertainties in one location may help at least to lessen uncertainties elsewhere. A strategy for biomass power must include ways to surmount these hurdles. One obvious option is to encourage integrated projects, in which all the interested parties participate on the basis of contractual commitments, as is common in more traditional energy developments. Integrated biomass power projects will become easier to establish when all the various stages of biomass production and utilization have been convincingly demonstrated.

One category of uncertainty is broadly common to all: the cost and performance of advanced biomass conversion technology – in particular close-coupled gasifier/gas-turbine technology, at the moment the most promising route to biomass power on a sizeable scale. The investment costs of the first generation of biomass gasifier/gas-turbine units – each the first of its kind, conservatively engineered and carrying the whole of the one-off development cost – is clearly too high to be economic, perhaps even with low-cost residue fuel and certainly with more expensive fuel from dedicated energy crops. Plant investment costs will come down, but how quickly? How much money will have to be spent to get up the 'learning curve' of engineering design? Current estimates, including for instance published figures from the Bahia project in Brazil,[66] suggest that the specific investment cost of a 'first-of-a-kind' prototype BIG-GT biomass power plant of 30 MW (electric) capacity might be of the order of US$2,500/kW, or a total capital cost of some US$75 million. The same estimates, and experience with the evolution of

---

[66] Elliott and Booth, *Brazilian Biomass Power Demonstration Project*.

other technologies, indicate that this cost could be reduced substantially over a series of similar units. By replicating a standard design, incorporating improvements based on earlier units, emphasizing shop fabrication of standard modules and rapid construction, and optimizing the mix of capital, operating and maintenance costs, a manufacturer could bring specific investment cost for this design down to US$1,500/kW over five plants, or US$1,300/kW over ten plants (see Box 6.1). Although still higher than the specific investment cost of a combined-cycle plant firing natural gas, this is beginning to look like a potentially commercially competitive technology. Moreover, since the plants are small, the total investment in ten plants, required to reach this stage of development, would be only some US$500–700 million. The financial support needed to move along this curve might start at, say, US$30 million for the first 30 MW unit, and be reduced progressively to zero over the first ten plants, in a decade or less. Once private investors see that such a learning curve is actually happening, private capital will become progressively more readily involved.

Until actual practical experience has been acquired, however, the total cost of biomass power at any particular location or with any particular technology can as yet be estimated only in broad-brush terms. Other cost factors that must be considered, depending on the organizational relationships in this early stage of development, might include land, labour, harvesting and other technology for biomass supply, plus storage and transport, with possible extras for comminution or forced drying. Taking the relevant factors together, and with allowance for profit for a fuel supplier, estimates indicate that the cost of biomass fuel from energy crops might be between US$1.5 and US$3/GJ at the power station gate.[67] The lower price would apply, for instance, in tropical or sub-tropical climates where land prices and labour costs are low; but in both industrial and developing countries, as field experience accumulates, a learning curve may likewise bring down at least some of the costs of producing and delivering biomass fuel. In general, for a 30 MW solid-fuelled power station, the cost of a unit of electricity sent out is typically about 60% capital recovery, with fuel cost and operating and maintenance costs making up the remainder. From the examples and assumptions in the study cited,[68] fuel at

[67] Ibid.
[68] Ibid.

US$2/GJ and a plant of 45% efficiency gives a fuel component of electricity cost of about 1.6 US cents/kWh, the capital-related component making up most of the rest, and of course depending on the rate of return required (see Box 6.1). Until prototype plants have operated for a reasonable period these estimates will remain simply estimates; but they look encouraging.

6.2 Biomass power in the EU

Consider, for example, a possible strategy for biomass power in the EU. Until recently, the emphasis of EU involvement in biomass energy has been on liquid fuels, particularly for transport. Comparatively little effort or funding has been devoted to options for advanced biomass electricity – and mainly to very small-scale activities such as work on ceramic turbines.[69] The probable evolution of biomass electricity will involve moving onwards beyond the present pattern. At the moment biomass, in the form of low-cost residue fuels including agriculture and forestry wastes and urban refuse, is burned directly as a boiler fuel, raising steam for a steam turbine, usually for cogeneration of electricity and heat, as described earlier. Expanding the role of biomass for electricity generation will entail:

- establishing appropriate energy crops grown explicitly as fuel for biomass power;
- demonstrating suitable technology for harvesting, drying, storing, transporting and preparing such biomass fuel; and
- developing and demonstrating conversion technology, most probably involving gasification and gas turbines, that can use the fuel efficiently and with minimal environmental impact to generate electricity and possibly heat.

Each of these three strands – crop selection and cultivation, biomass handling and processing, and biomass conversion to electricity – has already received significant attention both within the EU and elsewhere. The achievable aim of the next phase of RD&D within the EU could be to demonstrate integrated biomass power systems that incorporate optimum combinations of all three strands, appropriate to local growing conditions in various parts of the EU, and filling an appropriate niche in the local energy supply market.

---

[69] European Commission, DGXVII, *The European Renewable Energy Study.*

Box 6.1 Economics Overview

Electricity is a high value energy market, with wholesale prices typically ranging around five cents/kWh at the power plant. This is equivalent to \$14/GJ, \$85 per barrel of oil equivalent (boe), or around three times the wholesale price of automotive fuels in mid 1993.

Agricultural and forest industry residues are used to generate power in steam turbines generally below 25 megawatts electrical output (MW). In the USA, around 8 gigawatts (GW) of capacity operates in situations usually combining low feedstock costs with high electricity prices. Under the terms of the US Public Utilities Regulatory Policies Act (PURPA) of 1978, power utilities are obliged to buy electricity offered by independent generators at prices that reflect 'avoided costs' – the costs that would be incurred if the utility itself provided the additional power.

In the early 1980s, such costs often ran as high as nine cents per kWh and, at these guaranteed prices, there was a rush of developers to sign contracts.

However, as avoided cost levels dropped towards five cents per kWh, the flow of new biomass power projects has declined markedly because conventional steam cycle plants are handicapped by a combination of low efficiency and high specific investment cost at a scale suited to biomass applications.

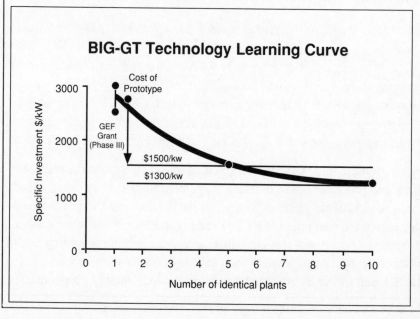

Recent assessments of emerging technologies suggest that power plants of a modest scale (20-50MW) could achieve thermal efficiencies in excess of 40 percent within a few years (eventually reaching 50 percent or more), combined with capital costs well below those of comparable conventional biomass plants utilising boiler/steam turbine technology.

A number of technological concepts are promising but the BIG-GT cycles are well-placed to make an early impact. This technology involves a gas turbine, closely coupled with an air-blown gasifier. Early plants are likely to incorporate a steam bottoming cycle (combined cycle), but other variants are possible such as an air bottoming cycle; steam-injected gas turbine (STIG); and intercooled, steam-injected gas turbine (ISTIG).

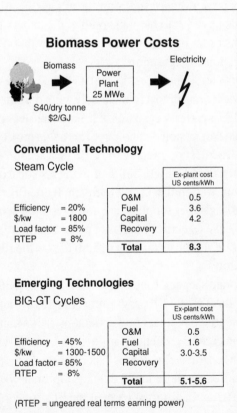

Biomass Power Costs

Biomass → Power Plant 25 MWe → Electricity

$40/dry tonne
$2/GJ

Conventional Technology
Steam Cycle

Efficiency = 20%
$/kw = 1800
Load factor = 85%
RTEP = 8%

	Ex-plant cost US cents/kWh
O&M	0.5
Fuel	3.6
Capital Recovery	4.2
Total	**8.3**

Emerging Technologies
BIG-GT Cycles

Efficiency = 45%
$/kw = 1300-1500
Load factor = 85%
RTEP = 8%

	Ex-plant cost US cents/kWh
O&M	0.5
Fuel	1.6
Capital Recovery	3.0-3.5
Total	**5.1-5.6**

(RTEP = ungeared real terms earning power)

Source: Brazilian Biomass Power Demonstration Project, Special Project Brief, P. Elliot and R. Booth, London, 1993.

What is needed now is a coherent programme within the EU, directed towards this objective.

Work on advanced biomass power outside the EU already making significant progress. As noted earlier, and as this is written, the Vaernamo unit in Sweden (6 MW electric and 9 MW district heating) is expected to supply its first biomass gas to the gas turbine before the summer of 1994; and ground has been broken for the 3 MW Paia biomass gasification power station in Maui, Hawaii, for startup in 1994. The EU does not need to reinvent the wheel. At the beginning of December 1993, applications were submitted to the European Commission for support from the THERMIE programme for the first biomass power demonstration unit in the EU. Although as this is written the details have yet to be made public, the understanding is that the unit may be of the order of 10 MW, and use short-rotation coppice fuel in a gasifier–cogeneration configuration.

The EU need not and should not stop at that. The next phase of EU development could get under way almost immediately, with the aim of demonstrating a fully integrated biomass power station at a more plausibly commercial scale, utilizing advanced conversion technology. The Bahia project now in progress in north-eastern Brazil, supported by the GEF (see section 4.5), offers a useful model on which the EU might build, drawing also on experience with the Puertollano demonstration IGCC unit in Spain. In the Brazilian project, two competing teams of gasifier suppliers are now preparing detailed engineering designs for a 30 MW (electric) biomass power station; the winning design will be chosen in 1994. The next phase of activities in the EU might therefore take shape as follows.

The EU Directorate-General for energy (DGXVII) announces, perhaps also in the context of the THERMIE programme, that it will support at least two different fully integrated biomass power demonstration units on a scale large enough to be commercial, in different parts of the EU, using different feedstocks and different conversion technologies – perhaps one in a northern member state and another in a southern member state. Optimal systems for biomass power will be different in different parts of the EU; supporting two different units will reduce the political infighting over choice of location, while fostering competition to reduce costs and achieve maximum performance. The different local climatic conditions will favour different local biomass crops.

Gasifier and gas turbine suppliers will compete to supply the conversion technology appropriate to the crops. The scale is important. Although a gasifier/diesel project may be economic at a scale of only 10–12 MW, BIG-GT technology is unlikely to be, because of the higher unit cost of turbine plant at such small sizes. If the EU is to build on the experience of the Bahia project in Brazil, the focus must be on commercial-scale units. Each unit should be of quasi-commercial size – at least 30 MW (electric). On the basis of the estimates cited above, total investment cost for two 30 MW prototype units would therefore be some US$140 million, or perhaps 100 million Ecu – that is, some 50 million Ecu per unit.

According to its rules the THERMIE programme can provide up to 40% of the costs – that is, some 20 million Ecu per unit; the remainder will come from participating organizations, including equipment suppliers, farming or forestry groups, host utilities, other utilities seeking access to the data generated, and possibly water companies. The Elcogas consortium building the Puertollano demonstration IGCC unit now includes some ten industrial organizations as equity participants. A similar cluster might form around each biomass gasification demonstration unit, at much lower total cost: whereas the 300 MW Puertollano unit will cost over US$860 million, or more than 500 million Ecu,[70] each biomass power demonstration unit, costing perhaps 50 million Ecu, could be funded by much more modest industrial contributions, with the EU providing the necessary additional support – no more than 20 million Ecu per unit, as noted. As with Puertollano, the European Commission might also act as a catalyst to assemble the relevant consortia. EU funding can be provided only for projects involving participation from at least two member states of the EU; however, despite the site-specific nature of biomass power, that should not be a problem for demonstration projects. Although both the fuel supply and the power station site will necessarily be located in the same member state, equipment suppliers and interested utilities from other states will almost certainly take an interest, as indeed may suppliers of fuel feedstocks from elsewhere in the EU, if only for testing purposes. Testing of fuels and components would be a comparatively minor expense –

---

[70] 'European IGCC Tilts at Windmills in La Mancha', *Modern Power Systems*, August 1992.

perhaps 5–10 million Ecu in total at most, depending on how much testing, and of which fuels and components, might be desirable.

Selection of the sites and the technologies will entail certain preliminaries, but much of the relevant work will already be in hand, in connection with the unit – understood to be at a scale of some 10 MW – for which THERMIE proposals were submitted in December 1993. Moreover, the Bahia project in Brazil will generate two techno-economic packages for 30 MW units based on Bioflow and TPS gasifiers. Much of the work will be directly applicable to 30 MW units in Europe if they are based on either Bioflow or TPS. The EU projects would save almost two years and some US$ 4 million by building on the Bahia project. For the EU projects the preliminaries will include:

- detailed examination of the productivity of selected energy-crop species under sustainable conditions, including field trials;
- trials of different harvesting, drying and other processing technologies;
- testing of biomass fuels in the pilot-plant gasifiers of the suppliers;
- detailed engineering design of each proposed gasifier and matching gas turbine, with suitable gas-cleaning stages; and
- in due course, construction and commissioning of the two demonstration units, integrating gasifier and gas turbine under operating conditions.

The projects must also comply with detailed environmental criteria, set out explicitly beforehand.

If the European Commission announced such a programme in 1994, a conservative timetable might allow at most a year for tendering, assessing proposals and preliminary testing, leading to a selection of sites and technologies before the end of 1995. A further 18 months to two years would suffice for detailed engineering design, at the same time establishing biomass fuel supplies. Such a schedule would also be timed appropriately for sequential implementation after the Brazilian project. Construction and commissioning of the EU units, beginning before the end of 1997, might take two to three years; and the units could be operating by the year 2000. This is, as indicated, a conservative timetable; if sufficient funds were available the demonstration units could be operating significantly sooner. Nor need a programme be limited just to two units; a case could be made for an entire series of units, designed in sequence, to take advantage of the 'learning curve' experience over perhaps five years, such that the later units would be nearing or indeed achieving full commercial status. 'Commercial status' will of course depend on local factors, such as a specific site, or the nature of the regional electricity system and its policies, for instance whether and at what price it may purchase electricity from independent generators. Ground rules like these already vary considerably from one member state to another; by the year 2000 the variations could be yet wider, or could be converging, depending on developments through the rest of the 1990s relating to privatization, independent generation, third-party access to grids, power 'wheeling' and other institutional arrangements, including possible initiatives and directives from Brussels.

Among the specific areas still requiring RD&D are the following:

- field trials of crop cultivation and harvesting, specifically for biomass power applications;
- drying and comminution of different crop materials to specifications for processing and feeding to various gasifiers;
- storage as a stage of processing for the same application;
- advanced feeding to pressurized reactors;
- gasification testing of processed crop materials;
- gas cleanup for different feedstocks; and
- gas turbine performance with fuel gas from different gasifiers and gasifier feedstocks, especially in STIG and eventually ISTIG configurations.

Costs will depend on how much testing is carried out and where: for instance, whether some tests require investment in pilot or testing facilities, or whether equipment suppliers already have suitable test rigs. As in the Bahia project in Brazil, the EU programme 'learning curve' should push 'value engineering' to the limit – 'the process of choosing solutions to specific technical problems in order to achieve some "optimal" mix of capital cost, reliability, operating cost, maintenance cost, energy consumption and so forth'.[71]

---

[71] Elliott and Booth, *Brazilian Biomass Power Demonstration Project*.

6.3 Other development opportunities

Although the specimen programme outlined above has been put forward in the context of the EU, similar programmes, suitably modified and with appropriate government or international support, could get under way elsewhere, for instance in various parts of the US, Canada, and non-EU Scandinavia, and indeed in Brazil. The National Biomass Power Program of the US DOE has put forward ambitious projections for expansion; but as yet it is still at an early stage. Although the US DOE and EPRI initiated in the spring of 1993 a bidding process for biomass power projects, little evidence is yet available about the response to the process; as this is written some 25 proposals have been received, but project selection will not be completed until well into 1994. Judging the speed and scale of eventual development of the US National Biomass Power Program at this stage would be premature; but the commitment appears convincing.

The potential for biomass power, using indigenous resources and offering major environmental benefits, could also be substantial in some of the countries of central Europe and the former Soviet Union; but fundamental problems of land use and land tenure would have to be resolved and stabilized before any serious promotion of energy crops as a quasi-commercial undertaking, for whatever purpose, could be undertaken. The geographical areas with perhaps the most important potential for biomass power are those in certain low-latitude countries still called, for lack of a more accurate label, developing countries. Electricity use in these countries is increasing rapidly, and indeed is often constrained by lack of generating capacity. Many developing countries can offer excellent conditions for fast-growing energy crops – as witness the number of countries long since major producers of sugar cane. The Bahia project in northeastern Brazil is of course designed to investigate and to demonstrate how biomass power might create a significant new option for electricity supply in developing countries. However, because biomass power may be significantly site-specific, other projects or programmes in other locations would be worth exploring. Elsewhere in Latin America candidate sites could certainly be found, as they could in some parts of Asia and Africa. But African sites might be constrained by lack of infrastructure, and some Asian sites might need to give priority to land – and, perhaps more importantly, water – for food production.

In any of these areas, in any case, an overriding requirement should be to ensure that biomass fuel is produced and harvested sustainably, perhaps from land already degraded by the destruction of existing forests.[72] The concept of biomass power could be fatally compromised if it were to be associated with further destruction of standing forests. For this and other related reasons, biomass power activities in developing countries will have to be organized and managed with special sensitivity, and may nevertheless prove controversial.[73] Although biomass power could come to play a major role in developing countries, the case remains strong for establishing it first as a valid electricity supply option in industrial countries – not least to prove that it is a modern technology, not a poor relation. In industrial countries, moreover, the influential agricultural lobby can give biomass power valuable momentum.

6.4 Emerging markets and sourcing for biomass power

Biomass power technology can expand along a number of different avenues. As noted earlier, some proponents, notably in the US but also elsewhere, are studying so-called 'co-firing' of biomass with coal. Co-firing entails feeding suitably prepared biomass into an existing coal-fired power station, using the biomass to replace a proportion of the coal that the station would normally burn. Co-firing biomass may involve adjusting the plant's fuel feed mechanisms and burners, both to handle the biomass fuel, the physical characteristics of which differ from those of coal, and to compensate for the lower energy density of the biomass. However, these disadvantages may be offset in particular by the extremely low or zero sulphur content of the biomass. For an older coal-fired plant facing expensive retrofitting or indeed shutdown because of sulphur-emission limits, the co-firing option could be a comparatively inexpensive lifeline. At the same time, switching a coal-fired unit to biomass co-firing could help to create a local market for biomass fuel crops, perhaps as a first step on the way to fully fledged biomass power generation in the locality. Indeed, as noted earlier, this could be an ideal way to initiate a

---

[72] Hall et al., 'Biomass for Energy'.
[73] Ibid. Examples of unsuccessful biomass projects in developing countries include the Jari project in Brazil and the Dendrothermal Power Program in the Philippines, both described here.

biomass power project. Co-firing biomass in an existing coal-fired station would create a firm biomass fuel market for farmers, without subjecting them to the pressure to produce the large volumes needed for a stand-alone biomass power station. Once biomass fuel production was fully established and proven, the fuel could be switched to a BIG-GT unit built on the same site. Although at the moment this option is being canvassed mainly in the US, it could be applied much more widely – for instance, in the UK or Denmark or any other region where coal-fired power is widespread and biomass fuel can be grown, possibly also including some parts of central and eastern Europe.

Another avenue often advocated is the route that begins with residue fuel – agriculture or forestry wastes – as a forerunner of dedicated energy crops. The advantage is obvious, in that the cost of electricity from a plant using residue fuel will be lower than that from a plant using fuel from a dedicated energy crop grown to earn revenue. Residue fuels may also, however, have disadvantages. The supply may be limited or unpredictable; and the physical and chemical characteristics of the fuel may also be unpredictable, possibly over a wide range. Drying and storing residue fuels may pose problems. Power-station equipment for residue fuel may not be appropriate for energy-crop fuel. Some residue fuels will be unsuitable for advanced generating technology like gas-turbines. If the longer-term intention is to launch fully fledged biomass power generation using energy-crop fuel, starting with residue fuel, if badly handled, may engender a negative public image that could damage the acceptability of the plant. Promoters can argue with justice that a suitable biomass power station could help to mitigate waste disposal and pollution problems, not aggravate them. Open, scrupulous and honest dialogue between the project and the local public, right from the initial planning stages, may avoid problems; but anyone contemplating this approach to biomass power should be aware of its possibly deleterious undertones, not only for the particular project in question but for biomass power as a modern high-technology energy supply option.

In most industrial countries electricity use is increasing only slowly,[74] and no major expansion of power generating capacity is anticipated. However – except in the unique and untypical case of the UK after electricity privatiza-

---

[74] *Electricity Supply in the OECD*, Paris: OECD International Energy Agency, 1992.

tion – the last major surge of orders for new capacity occurred in the late 1960s and early 1970s. By the turn of the century or soon thereafter the majority of these power stations will be 30 years old, and earmarked either for major refurbishment or shutdown. Among them are the coal-fired and nuclear power stations that now provide a substantial fraction of electricity supply. How will these stations be refurbished or replaced? At the moment the technology most favoured is clearly combined-cycle stations firing natural gas. But doubts remain about the price and reliability of natural gas supplies; if the necessary infrastructure is to be built, natural gas prices will have to rise substantially to fund the investment. Among the alternatives on offer to refurbish or replace existing power station capacity, biomass power may offer a distinctive group of advantages.

A mature biomass power station will have attributes much like those of an old traditional coal-fired station of similar output, or indeed of many modern 'clean coal' stations, with an output of 20–100 MW. It will be on a roughly similar physical scale, with similar requirements for site land. (It will, however, be much smaller and require much less land than a coal-fired station built in the 1960s or 1970s, using pulverized-coal technology in units of 500 MW or more.) To traditional electricity supply planners and engineers a mature biomass power station will look and behave much like a conventional power station – which other renewable electricity supply technologies like wind power still do not. It may be able to supply a significant electricity output, perhaps up to 100 MW, or possibly even more in certain places, and fit naturally into a traditional electricity despatching regime. But a biomass power station will emit almost no sulphur oxides, its NO_x emissions will be comparable to those from the best fossil-fired stations, and its carbon dioxide output will merely return to the atmosphere the carbon taken from it by the growing biomass fuel. Electricity suppliers in industrial countries are already looking ahead to the refurbishment or retirement of existing plant. These existing sites already have planning permission, water supply, facilities for fuel transport and storage, and connection to the local electricity grid. Because a biomass power station so closely resembles a coal-fired power station, at least some of these existing sites could be candidate sites for biomass power – provided that a suitable supply of biomass fuel can be grown within acceptable transport distances, and of course provided that all the relevant technologies have been shown to operate as intended and meet economic and environmental criteria. If these conditions are met, the market opportunities for rapid expansion of biomass power even in industrial countries could be considerable. A coal-fired unit being retired may of course be on a scale of 500 MW or more; it could not be replaced completely with biomass power at that site, because of the usual constraints of transport costs for the necessary biomass fuel. But the trend in electricity generation is towards smaller units in any case; and the site of a large pulverized-coal station could accommodate not only a biomass power unit but also, say, a natural gas combined-cycle plant or a 'clean coal' plant, both of which share with biomass power the advantages of smaller scale, modularity and rapid construction.

Whereas in industrial countries electricity use is increasing only slowly, in many developing countries it is now expanding rapidly; indeed, the rate of expansion of use is often constrained by lack of adequate generating capacity.[75] Electricity suppliers in these countries are now ordering and constructing new power stations as fast as they can, subject to limitations on available finance. In a number of developing countries, however, the ambitious expansion plans bring with them worrying environmental implications. International funding agencies like the World Bank are reluctant to finance nuclear power stations; but they have funded many traditional coal-fired and hydroelectric stations, whose environmental side-effects are now giving rise to serious concern. If mature biomass power technology could be successfully demonstrated it would offer an attractive alternative, with attributes especially apt for many developing countries.

A single BIG-GT unit, for instance, which might have a capacity in the range of 20–100 MW or perhaps larger in certain locations, would still fit comfortably into a comparatively small grid system, and could be ordered and brought on-stream quickly. Most of the hardware could be fabricated in factories and shipped to the site, reducing the problems associated with site fabrication, especially in locations that might entail difficult working conditions. Fuelling and operating an array of biomass power stations in rural areas would create a range of different long-term jobs, both skilled and semi-skilled, that could help to reduce the pressure on rural populations to leave

---

[75] World Energy Council, *Energy for Tomorrow's World.*

such areas for the overcrowded cities. The biomass fuel itself would be an indigenous resource, reducing the energy imports and debt-servicing problems facing many developing countries. Moreover, even a single biomass power station would be a project on a scale sufficient to justify the administrative overheads of an international funding agency like the World Bank; a series of replicated stations would be better still.

In both industrial and developing countries, therefore, the availability of biomass power as a mature integrated technology would offer a valuable option with substantial advantages. If biomass power is to reach this level of acceptance, it must move along the learning curve, with the requisite information campaigns and financial support, in order to reduce costs, demonstrate the desired technical and environmental performance, and persuade its many potential beneficiaries to take it seriously. Both the co-firing and the residue-fuel approach to biomass power can be and indeed are being set in train by private industry and private developers; the technical, financial and organizational risks are manageable and the anticipated returns reasonably well assured. But they are not thus far using advanced conversion technology. The integrated system approach, beginning with dedicated energy crops and proceeding through harvesting, processing, storage and use for electricity generation in an advanced-technology power station, still entails a suite of risks that may make private developers wary. If the integrated approach, with its much greater long-term potential, is to be established, it will need to be fostered, at least in its early stages, by active participation, both financial and organizational, from governments and international agencies. Fortunately such support is already emerging, as indicated above, in activities involving the European Commission and European national governments, the US DOE, and the GEF administered by the World Bank. But the potential of integrated advanced biomass power appears to warrant much more substantial promotional effort at the governmental and international level, in the near future, in both industrial and developing countries.

Biomass Power: Take-off?

7.1 When, where and how

Even with the most vigorous promotional support from national governments and international agencies, advanced biomass power is unlikely to prove itself technically mature and economically competitive until after the year 2000. It must first demonstrate that it has convincing answers to questions about the sustainable and environmentally acceptable production and processing of energy crops and about the conversion technology used to generate electricity. It must also show that the many players involved can establish suitable relations between them, on a normal commercial basis, with minimal if any participation from governments, as they routinely do in other commercial and industrial activities. Assuming that these issues can be satisfactorily resolved, the economic status of biomass power will remain to some extent site-specific, as already noted. In particular it will depend on the price that the local electricity system is prepared to pay for the output from a biomass power station. In a few temperate and high-latitude localities the cogeneration option will also be available; station operators can sell not only electricity but heat for district heating, at least during the winter heating season. But the lack of heat load at other seasons will bring down the capacity factor and worsen the economics of the unit. Industrial cogeneration with a year-round heat load may be more important, perhaps especially in tropical and subtropical countries with high biomass productivity and new industries. Nevertheless, the overriding determinant of the economic status of biomass power will be the price of electricity from competing technologies – especially natural gas fired combined-cycles and cogeneration plants. In the mid-1990s indications are that the cost of electricity from a BIG-GT unit firing energy-crop fuel could eventually be comparable with that of electricity from a natural gas fired unit; but the BIG-GT unit is unlikely to be able actually to undercut the natural gas unit unless the price of natural gas rises significantly. Whether this will happen is impossible to predict with any degree of certainty: but it is clearly a possibility in many candidate locations for BIG-GT, especially if natural gas prices rise to cover investment in the extensive new transmission infrastructure desired. Moreover, the largest markets for new generating capacity will be in developing countries. Many of these do not at the moment have access to natural gas for electricity generation, whereas many do have access, at least potentially and in some cases actually, to biomass fuels suitable for biomass power.

At one extreme, biomass power could remain of only peripheral interest, utilized on an ad hoc basis where residue fuels are available, and employing only traditional steam-cycle generating technology. At the other extreme, by 2050 it could be a major contributor to electricity generation in both industrial and developing countries, supplying baseload electricity from a widespread network of small and medium-sized power stations fuelled by energy crops on a straightforward commercial basis. Of the two extremes, the former seems unlikely, given the present scale of biomass power RD&D, and the burgeoning commitment of governments, international agencies and major industries; but the latter would come about only if concerns about carbon dioxide compelled governments to impose stringent restrictions on the use of fossil fuels, including natural gas. Otherwise the very availability of biomass power, not to mention other renewable electricity technologies like wind and possibly also photovoltaics, could suppress the demand for fossil fuels for electricity generation, perhaps even depressing fossil-fuel prices. A biomass power station based on gasification or some other conversion technology will always have a higher capital cost than an equivalent natural gas fired unit, simply because the biomass power unit must incorporate not only the gas-turbine generator but also the biomass conversion stage – the gasifier, or possibly eventually pyrolyzer, or some other intermediate stage. Accordingly, without some form of government intervention like controls or taxes on emissions of fossil carbon, biomass power will always be competing with fossil-fuelled alternatives whose prices are unlikely to be much higher and may be lower – provided, of course, that coal or natural gas is actually available in the given locality.

With these provisos, then, how might advanced biomass power be deployed, where, and how rapidly? Projections of possible energy futures tend to treat renewable energies as a single category. They rarely differentiate between renewables used to produce liquid fuels for transport and renewables used to generate electricity; and even when they do they often fail to distinguish sub-categories within renewable electricity technologies. For some examples of projections focusing specifically on biomass power, see Figure 7.1.

As already indicated, biomass power has more in common with traditional coal-fired generation than it has with other renewables like wind power and photovoltaics. The trajectory of deployment of biomass power could likewise be closer to that of coal-fired power than to those of wind and PV. Those with an active interest in seeing biomass power take off include, at the outset, immediate beneficiaries. Potential biomass fuel growers and the unions that represent their employees will see a new source of income. Agricultural departments and agencies will see an opportunity to reduce food-crop surpluses and subsidies and relieve the tensions they are now causing. Environment departments and agencies will see an opportunity to take practical steps towards reducing emissions of fossil carbon. Regional and development agencies will see a new local economic activity to provide local jobs and revenue to reinforce local economies in rural areas in both industrial and developing countries. Others may benefit in the longer term, as indicated in Chapter 2, but are unlikely to offer much support in the early stages. The initiative to promote biomass power must therefore come from governments and international agencies, with the active support of the biomass growers. Indeed, these growers – especially farmers and their organizations – might be the source of the strongest pressure at the outset, with their powerful political muscle, especially in the EU and the US.

Governments can prime the pump by offering bridging finance to the first tranche of biomass power stations; this in turn will help to create a market for energy crops. The crops themselves could benefit, at least in the early stages, from the efforts to help farmers adjust to changes in subsidies now given to food crops. Such a mechanism is already active, for instance, in member countries of the EU. Farmers can receive an EU payment for land 'set aside' and not used to grow food; this set-aside land, however, can be planted with an energy crop for eventual sale. Of course, energy crop plantations need not be confined to set-aside land; other land can also be used, and may qualify for

Figure 7.1 Projections of average electricity demand in biomass scenarios, with comparison of installed capacity over time in each

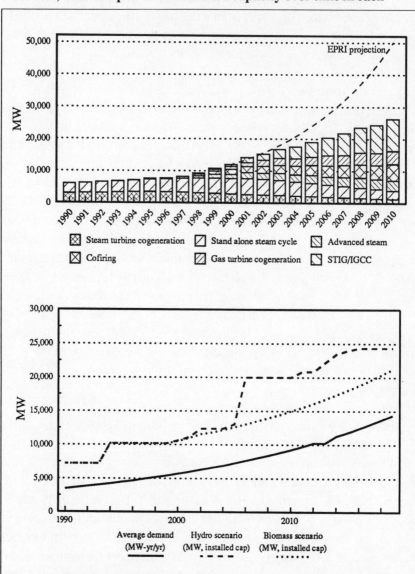

Sources: *Electricity from Biomass: National Biomass Power Program Five-Year Plan*, US Department of Energy, 1993; and 'Future Biomass-based Electricity Supply in Northeast Brazil', by A.E. Carpentieri et al., *Biomass and Bioenergy*, Vol. 4 No. 3, 1993.

other financial supports. The EU offers grants supporting forestry; and the UK government, for instance, also offers a Woodland Grant Scheme for planting woodlands, at least some of which might be earmarked as prospective energy crops. In this way the EU and member-state governments are already starting to act as financial buffers, encouraging the establishment of energy crops that will then be available for sale as biomass fuel, either for traditional applications or, in due course, for advanced biomass power stations.

If energy crops are already growing, potential power-station builders will become more committed to the option, especially if they too qualify for some financial support in the early years. In the UK, for instance, the Non-Fossil Fuel Obligation (NFFO) is already functioning as a financial support mechanism for renewable energy technologies. Biomass power in the UK will soon be a beneficiary of the mechanism. NFFO support payments are made on the basis of units of electricity produced; this applies pressure for greater efficiency and higher performance. To put this expenditure in context, any subsidies offered to help establish biomass power in the EU will be trifling compared to the existing framework of subsidies for food production.

Similar financial arrangements may be implemented in many other parts of the world, each of course structured and administered in the appropriate local context. Because agricultural production of any kind is already so heavily subject to government involvement virtually everywhere in the world, extending the framework to encompass biomass fuel production for power generation would not require major institutional innovations. The Brazilian government, for instance, already supports production of fuel ethanol for motor vehicles from sugar cane; and recent reports indicate that the state of São Paulo is now encouraging sugar refineries to generate electricity for supply to the grid, at a guaranteed price.[76] A further possibility would be to create new national or international agencies specifically dedicated to fostering renewable energy sources. Some commentators have called for a global Renewable Energy Agency, perhaps as part of the UN system, acting to promote renewable energy as the UN International Atomic Energy Agency has acted to promote nuclear power. Others have suggested a European Renewable Energy Agency, with a brief like that of the old European Coal and Steel

---

[76] 'State Woos Sugar Refineries', *Modern Power Systems*, December 1993.

Community. At first sight these proposals might be attractive; but if they are to be taken seriously they will have to be thought through with care, to avoid the bureaucratic problems all too endemic in international institutions.

In any event, if the initial support for demonstrations can show that biomass power is a technically feasible, economically promising and environmentally acceptable electricity supply option, the other potential beneficiaries identified above will be attracted to it, and add their momentum to its development. Consider, then, the corollaries of some of the projections suggested above. If the US DOE, EPRI, Brazilian and other projections for biomass power were achieved on the timetables stated, what effects would they have on key issues of importance? How would these implications be altered if the expansion of biomass power in these geographical areas were to be paralleled by similar expansion worldwide?

Some of the more obvious implications would include the consequent effect on:

• emissions of fossil carbon;
• land use;
• agriculture and trade;
• energy dependence, especially for developing countries;
• coal markets; and
• industrial markets, including strategic EU interests and US technology exports.

The site-specific aspects of biomass power make generalizations difficult and of limited value; the detailed effects will of course depend on the choice of particular fuels and technologies, the scale of individual units and systems, the nature of local electricity markets, local agricultural arrangements, and the prices and availability of competing fuels. But some broad-brush implications can already be sketched.

7.2 Fossil carbon emissions

Consider, for example, the US DOE National Biomass Power Program, and the EPRI projection that as much as 50,000 MW (electric) of biomass power could be in operation by the year 2010. This would certainly be ambitious –

some would say unrealistic; but for the purpose of analysing impacts the date of operation is less important than the capacity. If and when so much biomass power capacity is in operation, the implications will be considerable, on all the issues listed above. The total electricity generating capacity of the US in 1990 was 690,000 MW, expected to rise to 707,000 MW by 1995.[77] The target for biomass power is thus only some 7% of current capacity; if this much biomass power were displacing conventional coal-fired generating plant, it would eliminate about 90 million tonnes of fossil carbon emissions per year; and indeed, some years sooner, the growing biomass fuel would already be absorbing carbon dioxide from the atmosphere.[78] In 1990 US emissions of fossil carbon were 1,367 million tonnes.[79] In other words, so much biomass power would eliminate about 8% of current US emissions of fossil carbon. The assumption is oversimplified – it assumes, for instance, that both the biomass power plants and the coal-fired plants they displace are operating on baseload with 100% load factors; but the implication is nevertheless striking, especially if it is extended to other countries on a pro rata basis. Even on this oversimplified basis, biomass power could contribute usefully to meeting the longer-term targets implied by the Framework Convention on Climate Change. To be sure, this broad-brush extrapolation must be refined to take account of the other changes in electricity systems that will have happened by then and will still be happening; and the potential role of biomass power in different countries and on different systems will, as emphasized earlier, be to some extent site-specific and dependent on local conditions. Nevertheless the broad-brush numbers make clear that vigorous expansion of biomass power, in appropriate contexts and with appropriate arrangements, could play a significant role in global efforts to mitigate climate change.

Moreover, because biomass absorbs carbon dioxide while it is growing, prospective energy crops function as 'carbon sinks', as mentioned in the Convention. Some companies building fossil fuel fired power stations have already undertaken to plant forests on a scale sufficient to absorb as much

---

[77] OECD, *Electricity Supply in OECD Countries*.
[78] On the basis of a 1000 MW coal fired unit burning 3 million tonnes a year of bituminous coal that is 60% carbon.
[79] *Energy, Economics and Climate Change*, Cutter Information Corporation, Arlington, Massachussetts: October 1993.

atmospheric carbon as the stations emit, as so-called 'carbon offsets'. Strictly speaking, such absorption is temporary, on the geological timescale associated with the carbon in fossil fuels; even if the 'offset' trees are left standing they will die and decay within a century, releasing the carbon again. The trees will, however, constitute a sort of 'buffer store' for carbon, allowing time for other changes to occur – in particular, the possibility that when the trees are harvested they will be used to fuel biomass power stations that have replaced fossil fuel fired power stations in the interim. In this connection, an intriguing policy proposal has recently been put forward for what would be called a 'tradeable absorption obligation' or TAO.[80] In essence, the proposal is that 'energy sellers, at the wholesale level, are required to absorb some proportion of the carbon that is emitted when their product is used by the purchaser, or to contract with other firms to carry out this obligation'.[81] If implemented as a regulatory requirement, the TAO would establish an increasing resource base of potential biomass fuel, that could in due course replace an increasing fraction of fossil fuel. No TAO proposal could be implemented without encountering stubborn opposition in some quarters; nevertheless the underlying concept deserves closer consideration.

7.3 Land use

Still using the EPRI projection as a premise, consider the effect on land-use. According to the US DOE, a 150 MW advanced biomass power station would require energy crops grown on 100 square miles or 25,000 hectares of land. Once again, this is a broad-brush number; but it follows that 50,000 MW of biomass power of the same general characteristics would require fuel from 33,000 square miles or over 8,300,000 hectares of land. This is nearly 1% of the total land area of the US – a disconcertingly large figure, considering all the competing demands on land. On this basis, biomass power in the US could reach an upper limit from land-use considerations that would ultimately constrain its contribution to well under 10% of total US electricity supply. In the EU similar considerations apply, with probably similar biomass productivities in tonnes of dry matter per hectare at best, complicated by the

[80] Peter Read, *Responding to Global Warning*, London: Zed Books, 1994.
[81] Ibid.

higher population density of the EU. In the EU the total area of arable land set aside compulsorily from cultivation under the CAP at the moment is 4.6 million hectares;[82] the potential available, including voluntary set aside and pastoral land, might be as much as 22 million hectares.[83] If, say, 10 million hectares of this land were brought back into productive use to grow energy crops for biomass power, it might be able to support up to 60,000 MW of generating capacity, compared to a total generating capacity of some 456,000 MW in 1991.[84] Note, however, that these numbers are based on the assumption that all biomass power is fuelled by energy crops. Since some will undoubtedly be fuelled by residues, the land-use impact will be reduced accordingly.

In developing countries, however, the outlook may be very different, with the possibility in particular of much higher biomass productivity per hectare. To be sure, different conditions may apply. Questions of land tenure, of land use and also of water use for food production, and of environmental impacts will be much more sensitive and probably controversial, and will have to be resolved case by case. In some developing countries, nevertheless, the effect of land use for biomass power will be more acceptable than the effect of land use for alternative forms of electricity generation like hydroelectricity, as the study of the Bahia project by Carpentieri et al shows.[85] In developing countries, total electricity supply and electricity use per capita are both still well below the level in OECD countries; the future role of biomass power could therefore be much more significant than it appears likely to be in OECD countries, especially if developing countries also move rapidly to install the most efficient available end-use technologies for electricity use.[86]

---

[82] STOA, *Energy and Biomass*.

[83] *Non-food Uses of Agricultural Produce in Europe*, Brussels: Club de Bruxelles, 1992.

[84] Energy in Europe: Annual Energy Review, Brussels: European Commission, April 1993.

[85] Carpentieri et al., 'Future Biomass-based Electricity Supply in Northeast Brazil'.

[86] Jose Goldemberg et al., *Energy for a Sustainable World*, New York: Wiley Eastern, New Delhi: 1987.

7.4 Agriculture and trade

Probably the single most powerful impetus towards expanding biomass power in industrial countries will be the opportunity to make productive use of cultivated land no longer needed for food production. As the figures given above for land use indicate, biomass power could alleviate the problem considerably. Cultivating and processing the energy crops could create jobs in rural areas throughout the EU. Similar considerations apply to US agriculture. To be sure, depending on the particular energy crops grown, some of this work might be seasonal; some thought would have to be given to optimum organization of cultivation and processing for maximum local benefit and stable employment conditions. Nevertheless, this use for cultivated land no longer required for food production would ease the pressure on the agricultural sectors of major producer countries, and accordingly make international agreement on terms of trade relating to food exports and imports more readily achievable. Biomass fuel itself – at least, that appropriate for biomass power – would probably not constitute a significant item in international trade, because of transport costs; however, as the electricity sector itself becomes increasingly international, and power flows across international borders become more generally accepted as an article of trade, the contribution of biomass power to the international trade in electricity might become significant.

Biomass power could benefit the agriculture and trade of developing countries both directly and indirectly. Since electricity is a high value-added product, biomass fuel producers in developing countries like Brazil could earn enhanced profits by generating and selling electricity as well as, say, sugar, or even grow biomass exclusively for electricity generation. Such economic activity would also provide jobs in rural areas, to alleviate the pressure on rural populations to move into the crowded cities of developing countries. Moreover, expansion of biomass power in industrial countries could help to reduce surplus production of food in these countries, and perhaps thereby ease to some extent the problems of access to markets in industrial countries for food products from developing countries.

7.5 Energy independence and security

For some countries, the possibility of using indigenous biomass as fuel for electricity generation could be a welcome alternative to imports of coal or natural gas. The indigenous biomass would be grown and processed as a domestic economy activity, with the labour costs paid to the domestic workforce in local currency, removing or at least reducing what for some countries is a burden on the balance of payments. Moreover, using an indigenous fuel could improve energy security, by reducing the possibility that imported supplies could be interrupted or subjected to unilateral price increases. This is unlikely to happen with coal supplies, because coal is available from so many exporting countries and coal transport is relatively unconstrained by infrastructure requirements; but it is relevant for many systems that remain heavily dependent upon oil for power generation, or for those with growing dependence on natural gas, which is delivered mainly by large-capacity pipelines from supplier to user with very limited alternative routing in case of interference. The potential contribution of biomass power to energy independence and energy security will vary greatly from country to country; but in some countries at least – especially the developing countries of Africa and Latin America, with high biomass productivity and low population density – it may grow to made an important contribution to energy independence and security.

7.6 Coal markets and industrial transitions

The competing fuel most affected by a vigorous expansion of biomass power would undoubtedly be coal. On the basis that a traditional coal-fired power station with a capacity of 1,000 MW (electric) burns some 3 million tonnes of coal a year, the EPRI projection of 50,000 MW of biomass power by 2010 could displace up to 150 million tonnes of coal a year from the US electricity generating market – close to 15% of the total coal sales in the US in the early 1990s. Similar effects could take place wherever biomass power is vigorously promoted. At first glance, the immediate consequence of such a loss of market would be to depress coal prices, and make coal-firing cheaper, slowing the expansion of biomass power. But two further considerations could complicate matters. One is that many coal producers around the world have already cut their profit margins almost to the bone; indeed, in the EU the remaining domestic coal producers continue to receive subsidies ranging upward from US$20 a tonne, without which they would be hard-pressed to survive. Rather than offering yet lower prices, some coal producers might simply go out of business, reducing global coal output accordingly. The other consideration is that if evidence reinforces concerns about the greenhouse effect, some form of national or international control on the emission of fossil carbon may yet materialize. Coal will almost certainly bear the brunt of any such control measures, if only because governments are reluctant to tackle the other major source of fossil carbon emissions, the private car. The expansion of biomass power would offer a way to generate electricity from solid fuel while making no net contribution to greenhouse emissions, apart from those from machinery for cultivation, processing and transport, which are small compared to those from the fuel itself. Biomass power would therefore represent a further threat to the future of the world coal industry; but this threat would be trifling compared to that from government controls on emissions of fossil carbon.

Given that the outlook for coal in industrial countries is far from rosy, one further intriguing possibility might also be borne in mind. As noted earlier, a number of projects have already been established involving planting of trees as 'carbon sinks' to absorb from the atmosphere as much carbon as is being released in new coal-burning facilities, to offset the additional emissions. Hitherto these offsets have been implemented by the owner-operators of the new coal-burning facilities whose emissions are being offset. Coal producers themselves might, however, undertake to establish such offsets by tree-planting, perhaps in conjunction with participating in the construction of new 'clean coal technology' facilities.[87] In due course coal producers might then find themselves in possession of biomass fuel resources suitable for use in biomass power plants akin to the 'clean coal' facilities with which they would by then be associated. 'Carbon credits' – a sort of inverse carbon tax – could be given for carbon stock sequestered by growing biomass. The resulting plantation could then maintain a constant stock of sequestered carbon while being

---

[87] Patterson, *Coal-use Technology.*

cultivated and harvested sustainably to produce a continuous flow of fuel for biomass power stations; and the carbon credits could provide economic leverage for building, say, BIG-GT units to use the biomass fuel. The coming decades might then see the coal industry metamorphosing gradually into a biomass power industry, with coal miners evolving into plantation managers. In effect, the industry would be acquiring the ability to use biomass without first having to leave it underground for millions of years, and bringing the industrial revolution full circle – from wood to coal and at last back to wood.

7.7 Industrial markets

For the coal industry, unless it were to take an imaginative long-term view like that outlined above, biomass power would be simply more bad news. For the power plant construction industry, however, it could be good news. After the euphoric expansion of electricity supply systems in industrial countries in the 1950s and 1960s, the power plant constructors have been in the doldrums for more than a decade, with excess construction capacity scrambling for very few orders. If pressure to relieve agricultural problems led to an expansion of biomass power like that suggested by the EPRI and DOE projections, not only in the US but in the EU and elsewhere, the construction companies could look forward to a surge of orders. Moreover, the orders would be for a large number of comparatively small units, rather than a small number of large units. Such an ordering programme would also be beneficial for the constructors and their employees because it would mean shop fabrication of replicated units on a continuing basis, with a stable workforce, rather than site fabrication with a transient workforce.

An increasing number of orders might also be in the international market. The DOE has declared explicitly that a major objective of its National Biomass Power Program is to give the US a world lead in the technology, anticipating that export of biomass power technology around the world will become a major business.[88] If other industrial countries within the OECD do not wish to concede this business to the US without a competitive struggle, their plant manufacturers will have to seek comparable government backing. The

---

[88] US DOE, *Electricity from Biomass.*

scramble to make biomass power a commercial reality in the international technology marketplace could even itself become acrimonious, with bilateral export credit agencies striving to outdo each other as they once did in the battle for nuclear station orders.

For biomass power, however, unlike nuclear power, the World Bank could become a key financial player in the world market. The World Bank has always declined to support nuclear power projects; but it has already looked with favour on biomass power, through the involvement of the GEF in the Bahia demonstration project. The World Bank has been heavily criticized for supporting power projects detrimental to the environment. Biomass power, properly designed and operated, might be a welcome addition to its techno-logical portfolio. Active support from the World Bank, in turn, would give a major boost to biomass power.

Conclusions: Growing Importance

Around the world, the use of electricity is increasing, and will continue to increase for the foreseeable future. More efficient end-use in industrial countries may slow the rate of increase, but the rest of the world wants and expects to use much more electricity than it does today. Unfortunately, however, traditional electricity supply technologies, based on fossil fuels, hydro-electricity and nuclear power, are already causing concern because of the environmental and other side-effects they bring in train. At the same time, the world electricity business is undergoing profound changes. Many historic assumptions are being questioned and discarded. Only two decades ago conventional wisdom took for granted that a bigger power station was always a better power station. In the 1990s, on the contrary, as described in Chapter 5, many electricity system planners now see greater advantages in smaller generating units that can be planned, constructed and commissioned in three years or less, rather than six years or more.

A lengthening catalogue of new generating technologies is now coming into use for such small, efficient and more environmentally acceptable power stations; the catalogue includes natural gas combined cycles, 'clean coal' technologies and a number of renewable energy technologies. All these technologies fit well into the evolving philosophy of electricity planners, with its trend away from very large, centralized power generation to smaller-scale, decentralized, dispersed and diversified generation. Among the renewable options, biomass power offers distinctive attractions. At the upper end of the feasible size-range, perhaps between 100 and 150 MW per unit, a biomass power station, for instance a BIG-GT unit, will be much like a traditional fossil-fuelled power station of similar size, but cleaner and probably more fuel-efficient. It thus represents a comparatively smooth transition from traditional to innovative technology for system planners and operators.

In industrial countries, growing and processing biomass fuel for power generation offers the opportunity to establish a substantial new economic activity in areas that now produce too much food. Growing energy crops for biomass power would give farmers an alternative form of income; and as and when biomass power becomes commercially competitive it might eventually help to reduce the onerous burden of agricultural subsidies. In some developing countries, biomass power might be yet more important. In these countries the substantial quantities of usable biomass residues and the high productivity of certain energy crops, used with efficient conversion technologies, could make biomass a much more significant indigenous energy resource. Using indigenous biomass fuel for biomass power would reduce energy imports, alleviating the burden of foreign debt these countries bear, while helping to meet their burgeoning demand for electricity. At the lower end of the scale range, biomass power units based perhaps on gasifiers and diesel engines could deliver the electricity that would transform rural life in developing countries, while giving local people responsibility for and control over their local power requirements in a way that centralized systems cannot do.

In both industrial and developing countries biomass power could reduce the environmental impact of electricity systems, provided that suitably stringent guidelines are observed at every stage, from cultivation of energy crops through processing, transport and storage to control of power station emissions and wastes. Properly organized and managed, a biomass power system can function sustainably and in harmony with the environment, both local and global. Biomass power can thereby help the governments of the world to meet their international commitments, such as those made at the Rio conference in 1992.

However, biomass power still has a long way to go if its potential is to be realized. If it is to become a practical electricity option on a substantial scale worldwide it will require a wide-ranging array of research, development and demonstration, much of which will have to be more or less site-specific. Topics for RD&D will include: investigation and improvement of energy-crop species, to enhance productivity while minimizing local environmental impacts; optimal cultivation techniques for energy crops; proof of processes for harvesting, processing, transport, storage and drying of biomass fuels; and development and demonstration of various conversion technologies, including gasification at low and high pressure, small-scale gasification, gas cleanup, and integration of various gasifiers with diesel engines and gas-turbines, among others. Costs of fuel and conversion technology must come down, and both crops and conversion technology must be proven. Prospects appear encouraging, but questions must still be answered. This RD&D will require support from governments and international organizations. Fortunately, this support is already emerging, on a broad front, locally, nationally and internationally; and it is accompanied by increasing involvement of industrial participants – those who will eventually determine whether and how biomass power arrives in the commercial marketplace.

As always in these matters, of course, larger budgets for RD&D would propel development more rapidly and across a broader front. In the case of biomass power a particular case can be made for augmenting spending: because of its site-specific attributes, development and demonstration will have to be conducted in a wide range of conditions to be fully convincing. Energy-crop productivities, and fuel characteristics and performance, will have to be tested and proven locally for each region of interest and for each crop species, especially as breeding research advances and new clones become available. In general, the structure and focus of RD&D for biomass power appear reasonable at this early stage of progress; but one strong recommendation can be made. At the moment, RD&D on 'biomass' tends to be aggregated together in the minds of planners, administrators and budget overseers, such that work on various types of liquid fuels and other applications is lumped together with that on electricity generation. These topics are, however, becoming ever more widely differentiated both technically and economically. As yet the popular perception of 'biomass' – among those public, political and media people who give the matter any thought – is that 'biomass' automatically implies uses like transport fuels, in which the energy balances and the competitive prospects are significantly less encouraging than those for electricity from biomass. Biomass power deserves to be identified as a concept in its own right, within a policy framework appropriate to its distinctive attributes.

As described in earlier chapters, several of the most promising lines of development of biomass power are now on the threshold of the second 'D' in RD&D – demonstration. In the case of biomass power, what must be demonstrated is not just the technology of the power station but the entire system, from fuel production in field or forest through to electricity output. What are needed are not demonstration units but demonstration systems. Such a demonstration system requires a much larger financial outlay than R&D, and is therefore a more risky undertaking. However, demonstration systems – starting with dedicated residues or energy crops, and using advanced conversion technologies, to prove the performance and economics of complete integrated systems in many different locations – will be crucial if biomass power is to achieve its potential as quickly as it might. Designing, organizing and establishing such demonstration systems should be the main focus of the next phase of biomass power development around the world. It will require concerted efforts to assemble suitable consortia, with the necessary range of skills, experience and financial backing. Those who might take the necessary initiative could include prospective beneficiaries like farmers and foresters, probably in local groups, seeking new outlets for their produce, and industries that now have access to biomass residues that could become dedicated byproduct fuels. Government departments responsible for energy and agriculture could play a part, to ensure that the electricity output from demonstration plants was given access to the electricity market on fair and reasonable terms of payment, possibly including some redirection of existing subsidies away from food production in favour of fuel production. Other interim financial support from governments could include tax breaks and grants or low-interest loans to both fuel producers and electricity producers, to cover the risks associated with demonstration systems.

Even in industrial countries a case can be made for involving international agencies like the World Bank, on the basis that demonstrating biomass power in industrial countries will help to foster its acceptance as a valid modern electricity supply option in developing countries. However, the site-specific attributes of biomass power suggest that demonstration systems sited in developing countries – at scales from baseload grid-connected installations in tens of megawatts right down to village-based systems of 100 kilowatts or less – should be given priority for international support. A series of replicated local biomass power demonstration systems throughout a rural region of a developing country like India or China would justify the administrative overheads of an organization like the World Bank, while offering more immediate local benefits and fewer detrimental effects than many of the large-scale electricity supply projects the Bank has supported in the past.

If biomass power in its many manifestations can demonstrate its potential convincingly, the possible beneficiaries listed in Chapter 2 above may take a more active interest and play a more active role. Governments in many parts of the world may come to see biomass power as a way to alleviate problems of agricultural surpluses, environmental consequences of traditional generation and shortages of electricity supply, especially in rural areas. They may encourage electrical utilities and independent power producers to weigh the advantages of biomass power against other possible generating technologies, both for retrofits and for new units. Governments will also have to ensure that electricity regulators guarantee biomass power fair access to electricity grids, at fair prices. Once farmers, foresters and other actual and potential biomass fuel growers can see a genuine market for such produce, they may be prepared to devote more effort, including longer-term effort, to growing energy crops. Their trade organizations and the unions to which their employees belong may also exercise important political influence to persuade governments to support biomass power activities. Water companies may team up with biomass fuel growers to use sewage sludges as nutrients for energy crops. Wider availability of proven energy-crop fuel supplies may reinforce the interest of prospective users of biomass fuel. Engineering and construction companies and equipment manufacturers may see new markets open for modular small-scale biomass power units in many parts of the world, offering series sales of technologies that can be shop-fabricated by a stable workforce on a continuing basis. Regional and development agencies may see in biomass power an economic activity that fits well into local patterns of employment and society, and could bring other local activities like light industry into regional and rural areas. Environmental organizations may come to see biomass power, established and managed according to acceptable and sustainable environmental criteria, as a comparatively benign way to deliver the electricity that society desires, while minimizing undesirable side-effects, both local and global. If biomass power can demonstrate that it is a reliable, economic and environmentally acceptable source of electricity, electricity users may come to take it for granted. When electricity users take biomass power for granted, it will be fully established.

Biomass power cannot yet be taken for granted anywhere. Too few people know about it, or recognize its potential benefits. What is needed most at this stage, therefore, is wider dissemination of information about biomass power in its various manifestations, to those who should be interested in it – many people in many places. Biomass power may be able to produce the most versatile and flexible energy carrier, electricity, with high efficiency and low environmental impact, at a competitive price, from a sustainable renewable resource that can be stored and used as desired. Biomass power systems may be able to deliver outputs small enough to supply individual villages or factories, or large enough to supply baseload grid electricity. In industrial countries, growing fuel for biomass power may help to alleviate the problem of surplus food production by offering an alternative economic activity for agricultural areas. In both industrial and developing countries it may help to restore marginal or degraded land. In developing countries it may offer an opportunity to expand electricity supply in both urban and rural areas, using indigenous fuel resources under local control, avoiding imports and reducing foreign debt.

In short, biomass power may succeed. Whether it does will depend on whether its many potential beneficiaries promote the measures necessary to foster it. With the appropriate cultivation techniques, conversion technologies and system organization, biomass fuel need not be considered 'second best', or 'poor man's oil'. In many parts of the world it could prove to be a valuable, high-quality energy resource in its own right – especially for generating electricity. Uncertainties remain; but the promise is obvious and substantial. Biomass power warrants growing support.